Global Governance of the Environment

This work is dedicated to my mum and dad,
Mehri and Badiollah Akhtarkhavari.

Global Governance of the Environment

Environmental Principles and Change in International Law and Politics

---

Afshin Akhtarkhavari

Senior Lecturer, Griffith Law School, Australia

Edward Elgar

Cheltenham, UK • Northampton, MA, USA

Published by
Edward Elgar Publishing Limited
The Lypiatts
15 Lansdown Road
Cheltenham
Glos GL50 2JA
UK

Edward Elgar Publishing, Inc.
William Pratt House
9 Dewey Court
Northampton
Massachusetts 01060
USA

A catalogue record for this book
is available from the British Library

Library of Congress Number: 2010926005

ISBN 978 1 84980 255 0

Typeset by Servis Filmsetting Ltd, Stockport, Cheshire
Printed and bound by MPG Books Group, UK

Contents

Foreword

International environmental law has been one of the fastest developing areas of international law in modern times. While there were some indications of a growing international legal consciousness about the environment in the early part of the twentieth century, it was not until after the formation of the United Nations that real opportunities arose for determined multilateral environmental initiatives. The 1972 Stockholm meeting was a pivotal event in that process which set in train a host of measures which effectively gave birth to international environmental law. The 1992 Rio conference resulting in the Rio Declaration and associated conventions on Climate Change and Biological Diversity further solidified those gains in the development of the law and continue to be influential to the present day.

However, notwithstanding these developments international environmental law lacks a foundational instrument which identifies the core concepts and principles, goals and objectives, of the field. Here a contrast can be made with the field of international human rights law, which can look to the 1948 Universal Declaration on Human Rights and the subsequent 1966 International Covenants as providing the foundation for what has become known as the 'International Bill of Rights'. Likewise, in the law of the sea the 1982 United Nations Convention on the Law of the Sea is a foundational instrument which outlines the extent of coastal states rights and responsibilities and creates a framework for global oceans management and governance.

In the absence then of any overarching international environmental legal instrument, international environmental law has had to look to principles as the basis upon which the legal regime has been built. This has proven to be challenging in a field of international law which is comparatively young, and where some of those principles as they have been tentatively put forward have been heavily contested. The 2009 Copenhagen Climate Change negotiations served to illustrate just how fragile global consensus is on some environmental principles, and while the collapse of that process may ultimately reflect more upon the politics of climate change than environmental principles, it does highlight the difficulty in precisely identifying the principles upon which contemporary international environmental law is founded.

What this suggests is that identification of international environmental principles is increasingly significant in terms of understanding not only the foundation upon which international environmental law is based, but also how that law is to be interpreted and developed into the future. The principles themselves have of course been subject to much analysis and debate and in the early 1990s many scholars in the field expended considerable effort in identifying precisely what those principles were. However, given the immaturity of the international environmental legal system at the time, it was difficult to always identify a consistency in the state practice and implementation of those fledgling principles.

This is where Akhtarkhavari makes an important contribution in discussing the role of social learning and how there must first be an institutional appreciation of the issues before proper operationalisation and implementation can take place. This is especially highlighted in the context of the International Court of Justice, which in its very first judgement in the *Corfu Channel Case* outlined principles of state responsibility which had significant consequences for the developing law on transboundary environmental harm. Since that time, the Court has grappled with its understanding of environmental principles as highlighted by its decisions in the *Danube Dam* case and more recently in its 2010 *Pulp Mills* decision.

However, the International Court of Justice was only one institution undergoing the social learning process when it came to international environmental principles. The United Nations through its Global Compact was another, as were State parties to International Maritime Organization negotiations on ocean dumping. Through these cases studies of how various international institutions and actors grappled with environmental principles, Akhtarkhavari highlights the challenges the international community has faced in coming to terms with these principles and the consequences they have at both the international and national level.

In the absence of an overarching global environmental treaty, environmental principles remain pivotal to the further development of international environmental law, but perhaps most crucially to the operationalisation and implementation of the existing law by the international community and individual State actors. They will prove fundamental in resolving the climate change debate, but also in reassessing some existing international regimes such as that dealing with whaling. A better appreciation of those principles and how they are institutionally adopted – at all levels – remains fundamental for the future success of the international environmental project.

13 May, 2010

Donald R. Rothwell, Professor of International Law,
ANU College of Law, Australian National University, Canberra

Acknowledgments

I have always been interested in concepts, ideas, and institutions that help to unite people in a meaningful, sustainable and just fashion. The international community is still emerging from and grappling with trends that confirm rivalry as the dominant operating matrix for most transnational actors. In this context, abstract and open-textured norms like environmental principles are often viewed as tools for various actors to exact in different circumstances. My intuitions however suggested that these principles were also effective in uniting people to consult and engage in a discourse that ultimately contributed to important developments for the governance of the natural environment and resource consumption. I started this project to try and confirm or disprove my intuitions.

This book is based on my PhD thesis which was accepted in 2009. Professor Donald Rothwell was my supervisor and a long time mentor, and has always provided me with wise and insightful comments, criticisms and moral support throughout my academic career. I sincerely appreciate his help and friendship. I have also been fortunate over the years to have a number of friends and colleagues who have shaped how I work with international law. In particular Professors Emeritus Ben Boer and Ivan Shearer, and Professor Martin Tsamenyi have played an important role in developing my early activities as a scholar. My association with the Socio-Legal Research Centre at the Griffith Law School has fuelled my passion for interdisciplinary studies, not only because of the excellent colleagues I have been fortunate to have but also the scholars who have visited us over the years from all around the world. I developed and strengthened my appreciation of constructivism in international relations during discussions with Alexander Wendt, who at that time was at the Department of Political Science at the University of Chicago.

Members of my immediate family have been very patient and supportive during the time it took for me to complete this work. The love and support of my wife, Nikki, and our two children, Lachlan and Zoe, was most valuable during the project. I cannot thank them enough for everything they did to accommodate the hours I spent away from them while completing this work.

Some of this work has already been published elsewhere. In particular chapters 6, 7 and 8 are respectively based on these publications:

- 'Power, Environmental Principles and the International Court of Justice' (2009) 28 *Australian Yearbook of International Law* 1.
- 'The Global Compact, Environmental Principles, and Change in International Environmental Politics' (2010) 38(2) *Denver Journal of International Law and Policy* 101.

Some of the material in Chapter 8 was (first) presented at the Centre for International and Public Law (CIPL) – Australian Centre for Environmental Law (ACEL) workshop, 13–15 August 2009, *Environmental Discourses in International and Public Law* at the Australian National University and is also being developed separately for Kim Rubenstein and Brad Jessup (eds) (2010), *Environmental Discourses in International and Public Law*, Cambridge: Cambridge University Press.

During the early stages of the editing process I received very useful support from Fiona Lubbett. Finally, the staff at Edward Elgar have been incredibly helpful during every stage of the production of this book. I sincerely thank them for their efficient, thorough and friendly assistance. In particular, I want to thank Georgiana Appelgate, Alex Pettifer and David Fairclough.

The law and web pages referred to are current as at January 2010.

Figures

Abbreviations

CEO	Chief Executive Officer
CO_2	Carbon Dioxide
COP	Communication on Progress
GC	Global Compact
GCLN	Global Compact Local Networks
GCO	Global Compact Office
GGE	Greenhouse Gas Emissions
ICESCR	*International Covenant on Economic, Social and Cultural Rights*, opened for signature 16 December 1966, 999 UNTS 3 (entered into force 3 January 1976).
ICJ	International Court of Justice
ILA	International Law Association
ILC	International Law Commission
ILM	International Legal Materials
ILO	International Labour Organisation
IMO	International Maritime Organisation
IPCC	Intergovernmental Panel on Climate Change
ITLOS	International Tribunal for the Law of the Sea
LC	*Convention on the Prevention of Marine Pollution by Dumping of Wastes and Other Materials*, opened for signature 29 December 1972, 1046 UNTS 120 (entered into force 30 August 1975).
LP	*Protocol to the Convention on the Prevention of Marine Pollution by Dumping of Wastes and Other Matter*, opened for signature 7 November 1996, (1997) 36 ILM 1 (entered into force 24 March 2006).
MEA	Multilateral Environmental Agreements
NGO	Non-governmental Organisation
OECD	Organisation for Economic Co-operation and Development
OHCHR	Office of the High Commission for Human Rights
TNC	Transnational Corporations
UN	United Nations
UN GA	United Nations General Assembly
UNCED	United Nations Conference on Environment and Development

UNCLOS	*United Nations Convention on the Law of the Sea*, opened for signature 10 December 1982, 1833 UNTS 397 (entered into force 16 November 1994)
UNDP	United Nations Development Programme
UNIDO	UN Industrial Development Organisation
UNODC	United Nations Office on Drugs and Crime
UNTS	United Nations Treaty Series
WTO	World Trade Organisation

Table of international cases

Table of international treaties and other international instruments

PART I

Introduction

1. Introduction

1.1 INTRODUCTION

Environmental principles are commonly used in international politics amongst a variety of actors including states, non-governmental organisations and multi-national corporations.[1] In an empirically based study of global business regulation, which extended to environmental issues, Braithwaite and Drahos suggested that a significant volume of negotiations internationally is carried out using principles.[2] Others like Kiss and Shelton write that '[p]rinciples are widely used in international environmental law, perhaps more than in any other field of international law'.[3] Sands has also suggested that the environmental principles in his list have received broad, 'if not necessarily universal, support and are frequently endorsed in practice'.[4]

Interestingly, and despite the obvious importance of environmental principles at the international level, there is as yet no instrument binding under international law which sets out the general principles of international environmental law and politics.[5] Historically, the Declaration that was

---

1 See for instance Nicolas de Sadeleer, *Environmental Principles: From Political Slogans to Legal Rules* (2005), who at 258 comments that environmental law is strongly 'marked by the presence of principles compared with other legal disciplines'.

2 John Braithwaite and Peter Drahos, *Global Business Regulation* (2000) 29.

3 Alexandre Kiss and Dinah Shelton, *International Environmental Law* (3rd ed, 2004) 203. Nagendra Singh, a former President of the International Court of Justice, went as far as to suggest that sustainable development as an environmental principle was a peremptory norm in international law which no treaty or customary practice could breach: see Nagendra Singh, 'Sustainable development as a principle of international law' in Paul De Waart, Paul Peters and Erik Denters (eds), *International Law and Development* (1988) 1–12; Nagendra Singh, 'Foreword' in World Commission on Environment and Development, R D Munro, and J G Lammers, *Environmental Protection and Sustainable Development: Legal Principles and Recommendations* (1987) 1–4.

4 Phillipe Sands, *Principles of International Environmental Law* (2nd ed, 2003) 231.

5 Patricia Birnie and Alan Boyle, *International Law and the Environment* (2nd ed, 2002) 21; David Hunter, James Salzman and Durwood Zaelke, *International*

signed at the Stockholm Conference on the Human Environment in 1972 was the first effort at the international level to use the language of environmental principles to refer to a series of norms listed into an international instrument;[6] the *Stockholm Declaration* contained 26 principles. It has been argued that the *Stockholm Declaration* was a turning point in terms of tackling environmental concerns at the international level.[7] In his 1972 speech to the General Assembly of the United Nations, Maurice Strong, who later became the first Executive Director of the United Nations Environment Programme, made the following observation highlighting the significance that the principles had, at that time, for those who had drafted it:

> It is the first acknowledgement by the community of nations of new principles of behaviour and responsibility which must govern their relationship in the environmental era. And it provides an indispensable basis for the establishment and elaboration of new codes of international law and conduct which will be required to give effect to the principles set out in the Declaration.[8]

Environmental Law and Policy (1998) 320–321; Sands, above n 4, 231–232; de Sadeleer, above n 1, 243.

[6] See the *Declaration of the United Nations Conference on the Human Environment*, UN Doc A/CONF.48/14/Rev.1 (1972) ('*Stockholm Declaration*'). Much has been written on the significance of the Stockholm Conference on the Human Environment; for a general introduction see, for instance, Lynton Caldwell, *International Environmental Policy* (3rd ed, 1996) especially chapters 2 and 3; Louis Sohn, 'The Stockholm Declaration on the Human Environment' (1973) 14 *Harvard International Law Journal* 423; Birnie and Boyle, above n 5, 37–45. The idea that the *Stockholm Declaration* codified and listed environmental principles for the first time is not to suggest that the ideas contained within them were used for the first time only in 1972; for instance, Principle 1 refers to intergenerational equity: Edith Brown Weiss (ed), *Environmental Change and International Law: New Challenges and Dimensions* (1992) 385–412; Edith Brown Weiss, *In Fairness to Future Generations: International Law, Common Patrimony, and Intergenerational Equity* (1989); Birnie and Boyle, above n 5, ch 3, especially section 2(2). However the ideas behind this principle were, for instance, mentioned as far back as 1946 in the *International Convention for the Regulation of Whaling* in what would now be a preamble to an international agreement; the Convention recognised 'the interest of the nations of the world in safeguarding for future generations the great natural resources represented by the whale stocks': *International Convention for the Regulation of Whaling*, opened for signature 2 December 1946, 161 UNTS 72 (entered into force 10 November 1948).

[7] For instance, Andronico O Adede, 'International Environmental Law from Stockholm to Rio – an Overview of Past Lessons and Future Challenges' (1992) 22(2) *Environmental Policy and Law* 88; Ranee Panjabi, 'From Stockholm To Rio: A Comparison of the Declaratory Principles of International Environmental Law' (1992–1993) 21 *Denver Journal of international Law and Policy* 215.

[8] Quoted in Sohn, above n 6, 431.

In some cases environmental principles contained in the *Stockholm Declaration* expanded on previous responses to the rights states had to their natural resources. For instance, Principle 21 of the *Stockholm Declaration* connected what had already been established during the 1960s as the sovereign right states had to their natural resources with the provision that the 'activities within their jurisdiction or control' should not 'cause damage to the environment of other States or of areas beyond the limits of national jurisdiction'.[9] The principles contained in the *Stockholm Declaration* arguably formed the foundations for a variety of subsequent normative developments at the international and domestic level. A number of the principles in the *Stockholm Declaration,* when combined, highlight the concern at that time with sustainability, which served as the basis for the later codification of the principle of sustainable development at the international level.[10] Principle 2 of the *Stockholm Declaration* referred to the rights of 'future generations' which is seen as the early conceptualisation of the intergenerational equity principle frequently referred to in international environmental law and politics.[11]

The initiative to list environmental principles of international significance has since been followed in other instruments such as the 1982 United Nations General Assembly Resolution entitled the *World Charter for Nature,*[12] and of greater importance, the 1992 *United Nations Declaration*

---

[9] See Principle 21 of the *Stockholm Declaration* which provides for 'the sovereign right' of states 'to exploit their own resources pursuant to their own environmental policies'. In 1962 the UNGA had passed resolution 1803(XVII) acknowledging the sovereignt right of states to their natural resources. This absolute right was limited in 1972 by Principle 21 of the *Stockholm Declaration* requiring that states take into account environmental concerns.

[10] Daniel Magraw and Lisa Hawke, 'Sustainable Development' in Daniel Bodansky, Jutta Brunnée and Ellen Hey (eds), *The Oxford Handbook of International Environmental Law* (2007) 613, 615. Magraw and Hawke, at 615, refer to Principles 1, 8, 11, 21 and 23 as the foundations for the development of the concept of sustainable development in international law. See also the statement by the UNGA when convening the Stockholm Conference on the Human Environment, where it wrote about 'an urgent need for intensified action, at national and international level to limit, and where possible, to eliminate the impairment of the human environment': *Problems of the Human Environment,* GA Res 2398, UN GAOR, 23rd sess, 1733rd plen mtg, UN Doc A/Res/2398 (1968).

[11] See Brown Weiss, *In Fairness to Future Generations*, above n 6, 24. On this point see also Elli Louka, *International Environmental Law: Fairness, Effectiveness, and World Order* (2006) 30.

[12] *World Charter for Nature*, UN GAOR, GA Res 37/7, 48th plen mtg, UN Doc A/RES/37/7 (1982). The *World Charter for Nature* lists 5 environmental principles but, unlike the *Stockholm Declaration,* the resolution contains other parts that spell out the scope of the principles in terms of their function and implementation.

on Environment and Development.[13] The *Rio Declaration* codified and listed 27 principles that framed the proceedings of the United Nations Conference on Environment and Development held in Rio de Janeiro in 1992.[14] Observers have noted that before the *Rio Declaration* it was rare for conventions or treaties to have separate articles in them devoted to environmental principles relevant to those agreements.[15] Since the United

---

The *World Charter for Nature* had significance in terms of future environmental agreements. For instance see the preamble to the *Basel Convention on the Control of Transboundary Movements of Hazardous Wastes and Their Disposal,* opened for signature 22 March 1989, 1673 UNTS 126 (entered into force 24 May 1992), which refers to the *World Charter for Nature* as being about the 'rule of ethics in respect of the protection of the human and the conservation of natural resources'.

13 UN Doc A/CONF.151/5/Rev.1 (1992) (*'Rio Declaration'*). For commentary on the *Rio Declaration* see, for instance, David Wirth, 'The Rio Declaration on Environment and Development: Two Steps Forward and One Back, or Vice Versa' (1995) 29 *Georgia Law Review* 599; Ileana Porras, 'The Rio Declaration: A New Basis for International Cooperation' in Philippe Sands (ed), *Greening International Law* (1993) 20; Marc Pallemaerts, 'International Environmental Law From Stockholm to Rio: Back to the Future?' in Philippe Sands (ed), *Greening International Law* (1993) 1.

14 Along with the *Rio Declaration* several other significant instruments came from the United Nations Conference on Environment and Development. They included *Agenda 21* UN Doc A/CONF.151/26/Rev.1 (1992); the *Non-Binding Principles on the Sustainable Development of all Types of Forest* (31 ILM 881 (1992)); *Convention on Biological Diversity*, opened for signature 5 June 1992, 1760 UNTS 143 (entered into force 29 December 1993); *United Nations Framework Convention on Climate Change,* opened for signature 9 May 1992, 1771 UNTS 165 (entered into force 21 March 1994). *Agenda 21*, which was designed to clarify the scope of the environmental principles in the *Rio Declaration*, also contained principles of its own right; see de Sadeleer, above n 1, 312. With respect to the impact of the United Nations Conference on Environment and Development upon the development of environmental norms see David Freestone, 'The Road from Rio: International Environmental Law after the Earth Summit' (1994) 6 *Journal of Environmental Law* 193.

15 Daniel Bodansky, 'The United Nations Framework Convention on Climate Change: A Commentary' (1993) 18 *Yale Journal of International Law* 451, who lists the following pre-1993 environmental agreements as containing clauses with environmental principles in them: *Convention on International Trade in Endangered Species of Wild Fauna and Flora*, opened for signature 3 March 1973, 993 UNTS 243 (entered into force 1 July 1975) ('*CITES*'); *Convention on the Conservation of Migratory Species of Wild Animals*, opened for signature 23 June 1979, (1980) 19 ILM 15 (entered into force 1 November 1983) (*'Bonn Convention'*); *Protocol on Environmental Protection to the Antarctic Treaty*, opened for signature 4 October 1991, (1991) 30 ILM 1461 (entered into force 14 January 1998). Interestingly, other important pre-UNCED agreements did not refer to environmental principles; see, for example, the *Vienna Convention for the Protection of the Ozone Layer*, opened for signature 22 March 1985, 1513 UNTS 324 (entered into force 22 September

Nations Conference on Environment and Development in June 1992, which resolved in favour of the *Rio Declaration,* many multilateral agreements, declarations and resolutions amongst states have included articles listing relevant environmental principles in them.[16]

A variety of international courts and tribunals have considered environmental issues and applied international environmental law to resolving disputes.[17] Several have referred to or decided cases drawing upon environmental principles.[18] The precautionary principle has been considered

---

1988) ('*Convention for the Protection of the Ozone Layer*') and the *Montreal Protocol on Substances that Deplete the Ozone Layer*, opened for signature 16 September 1987, 1522 UNTS 29 (entered into force 1 January 1989) ('*Montreal Protocol*'), neither of which refers to the precautionary principle in the substantive provisions of the agreement; instead in both agreements there is a reference in their preambles to the 'precautionary measure'.

[16] This is not to suggest that before then international agreements did not refer to environmental norms that were potentially open-textured or abstract. In fact, Koskenniemi lists and discuses a number of international environmental agreements before 1992 which contained normative provisions that fail to develop concrete measure for actors to implement: see Martti Koskenniemi, 'Peaceful Settlement of Environmental Disputes' (1991) 60 *Nordic Journal of International Law* 73. Some examples he gives include: art 2(1) of the *Convention on the Protection of the Ozone Layer,* opened for signature 22 March 1985, 1513 UNTS 324 (entered into force 22 September 1988); and arts 192 & 193 of the *United Nations Convention on the Law of the Sea*, opened for signature 10 December 1982, 1833 UNTS 397 (entered into force 16 November 1994) ('*UNCLOS*'). Examples of treaty provisions concluded in and after 1992 with explicit references to environmental principles include, for instance, art 3 of both the *United Nations Framework Convention on Climate Change*, opened for signature 9 May 1992, 1771 UNTS 165 (entered into force 21 March 1994), and the *Convention on Biological Diversity*, opened for signature 5 June 1992, 1760 UNTS 143 (entered into force 29 December 1993). Two agreements that include reference to, for instance, the precautionary principles are art 4 of the *Stockholm Convention on Persistent Organic Pollutants*, opened for signature 22 May 2001, 40 ILM 532 (entered into force 17 May 2004), and art 6(3) and (5) of the *International Convention on the Control of Harmful Anti-fouling Systems on Ships*, opened for signature 5 October 2001, 40 ILM 532 (2001) (entered into force 17 September 2008).

[17] For studies of environmental issues in international courts and tribunals see Ellen Hey, *Reflections on an International Environmental Court* (2000); Amedeo Postiglione, *The Global Environmental Crisis: The Need for an International Court of the Environment* (1996); Alfred Rest, 'The Indispensability of an International Environmental Court' (1998) 7 *Review of European Community and International Environmental Law* 63; and Tim Stephens, *International Courts and Environmental Protection* (2009).

[18] Scholarly works examining environmental principles in international dispute resolution are usually of two different kinds. They either assess how a specific principle has been applied by tribunals or they are part of a broader analysis of a

in decisions of the World Trade Organisation's (WTO) dispute settlement bodies,[19] as well as the International Tribunal for the Law of the Sea (ITLOS).[20] The International Court of Justice (ICJ) and the Appellate Body of the WTO have also referred to the principle of sustainable development.[21] Several international courts and tribunals have made references

---

particular decision. As an example, Philippe Sands in 'International Courts and the Application of the Concept of "Sustainable Development"' (1999) 3 *Max Planck Yearbook of United Nations Law* 389 discusses the principle of sustainable development in the context of the *Case Concerning the Gabčíkovo-Nagymaros Project (Hungary/Slovakia)* [1997] ICJ Rep 7 ('*Danube Dam Case*') and *United States – Import Prohibition of Certain Shrimp and Shrimp Products; Recourse to Article 21.5* WTO Doc WT/DS58/AB/RW (2001) (Report of the Appellate Body) ('*Shrimp Turtle II*'). In contrast, Afshin A-Khavari and Donald Rothwell in 'The ICJ and the Danube Dam Case: A Missed Opportunity for International Environmental Law?' (1998) 22 *Melbourne University Law Review* 507 examine the decision in the *Danube Dam Case* and in that context explore the significance of the principle of sustainable development.

[19] See. e.g., *European Communities – Measures Concerning Meat and Meat Products (Hormones)* WTO Docs WT/DS26/R/USA, WTO Doc WT/DS48/4/CAN (Report of the Panel) WTO Doc WT/DS26/AB/R, WT/DS48/AB/R (1998) (Report of the Appellate Body) [123]; *European Communities – Measures Affecting the Approval and Marketing of Biotech Products (Biotech Products)* WTO Docs WT/DS291/R, WT/DS292/R, WT/DS293/R (2006) (Report of the Panel) [7.89]. The issue of concern in both these cases was whether the precautionary principle was a principle of international law. For the relevance of the principle for dispute settlement and the World Trade Organisation more generally see Jan McDonald, 'Tr(e)adding cautiously: precaution in WTO decision making' in Elizabeth Fisher, Judith Jones, Rene von Schomberg, *Implementing the Precautionary Principle: Perspective and Prospects* (2006) 160. For a more general overview of the precautionary principle or approach in the work of the World Trade Organisation dispute settlement see, de Sadeleer, above n 1, 103–108.

[20] The following cases decided by the ITLOS have dealt with environmental issues: *Southern Bluefin Tuna (Provisional Measures)* (1999) 38 ILM 1624 ('*Southern Bluefin Tuna*'); *MOX Plant (Provisional Measures)* (2001) ITLOS No.10; *Case concerning Land Reclamation by Singapore in and around the Straits of Johor (Malaysia* v. *Singapore)* (Provisional Measures, Order of 8 October 2003) ITLOS No. 12. The *Southern Bluefin Tuna* case at paras [77]–[79], is the only case that has given serious consideration to the precautionary approach or principle: see Alan Boyle, 'The Environmental Jurisprudence of the International Tribunal for the Law of the Sea' (2007) 22(3) *The International Journal of Marine and Coastal Law* 369, 373–376.

[21] The ICJ referred to the principle of sustainable development in the *Case Concerning the Gabčíkovo-Nagymaros Project (Hungary/Slovakia)* [1997] ICJ Rep 7, [140]. See chapter 6 of this book for a discussion of this principle in the context of the *Danube Dam Case*. The WTO Appellate Body in *Shrimp Turtle II* also discussed sustainable development in the context of Art XX(g) of the 1947 *General Agreement on Tariffs and Trade,* opened for signature 30 October 1947,

to and formulated versions of the basic notions found in Principle 21 of the *Stockholm Declaration* and Principle 2 of the *Rio Declaration* relating to the obligations states have to avoid causing harm to the environment of neighbouring states.[22] References to environmental principles are potentially even more common in written and oral pleadings before international courts and tribunals.[23]

At the international level, environmental principles have also been codified in contexts that have relevance for actors other than states. In particular, transnational corporations have expressed interest in environmental principles codified into institutions at the international level that have been set up by states.[24] Prominent examples of this include

---

55 UNTS 187 (in force provisionally under the *Protocol of Application,* 55 UNTS 308): *United States – Import Prohibition of Certain Shrimp and Shrimp Products; Recourse to Article 21.5* WTO Doc WT/DS58/AB/RW (2001) (Report of the Appellate Body), [125]–[145].

[22] For example, the ICJ in the following three cases discusses this principle: *Case Concerning Pulp Mills in the River Uruguay (Argentina v Uruguay), Request for the Indication of Provisional Measures,* Order of 13 July 2006, 45 ILM 1025, [72]; *Legality of the Threat or Use of Nuclear Weapons (Advisory Opinion)* [1996] ICJ Rep 226, [29]; *Case Concerning the Gabčíkovo-Nagymaros Project (Hungary/ Slovakia)* [1997] ICJ Rep 7, [140].

[23] See for example the numerous references to and use made of the precautionary principle in the 'Minutes of Public Sittings' of the *Southern Bluefin Tuna Cases* (*New Zealand* v. *Japan*; *Australia* v *Japan*), Provisional Measures (18 to 20 August 1999), www.itlos.org/start2_en.html. In the *Danube Dam Case*, the State of Hungary in its *Memorial of the Republic of Hungary in the Case Concerning the Gabčíkovo-Nagymaros Project (Hungary v Slovakia)* Vol 1 (2 May 1994), esp. Part III, www.icj-cij.org/docket/index.php?p1=3&p2=3&k=8d&case=92&code=hs&p3=1 also made wide ranging references to environmental principles, including the precautionary principles, sustainable development, and environmental impact assessment.

[24] At the United Nations, 'transnational' is the preferred term over 'multinational'; see Viljam Engström, *Realizing the Global Compact* (2002) 5, who argues that a transnational corporation distinguishes itself from other corporations by being able to 'locate production across national borders, to trade across frontiers, exploit foreign markets, and organize managerial structures in a way that affects the international allocation of resources'. See also Peter Muchlinski, *Multinational Enterprises and the Law* (1995); Luzius Wildhaber, 'Some Aspects of the Transnational Corporation in International Law' (1980) 27 *Netherlands International Law Review* 79. This discussion is based on the assumption that transnational corporations themselves choose how to engage with the principles prescribed for them at the international level. This is different, for instance, from something like the *Draft Norms of Responsibilities of Transnational Corporations and Other Business Enterprises with Regard to Human Rights*, E/CN.4/Sub.2/2003/12 (2003) which applies to transnational corporations after they have been adopted by the government of the country that they are incorporated in. Also, other actors

the Global Compact initiative of the United Nations,[25] and the *OECD Guidelines for Multinational Enterprises*.[26] The *OECD Guidelines for Multinational Enterprises* adopt provisions listing the expectations that member states have of transnational corporations.[27] In the commentary on the Guidelines, the OECD highlights the fact that the environmental provisions 'reflect the principles and objectives of the Rio Declaration'.[28] The Global Compact initiative of the United Nations is markedly different from the *OECD Guidelines for Multinational Enterprises* because it simply adopts three environmental principles to frame the learning initiatives developed within the regime.[29] In the case of both these initiatives, states have used environmental principles developed to guide states for the purposes of giving direction to the behaviour of transnational corporations.

This brief description of the significant use of environmental principles at the international level is not meant to suggest that they are all the same in terms of historical relevance, substantive content or normative persuasiveness.

---

are not bound internationally to environmental principles, whether it be through soft or hard law. For instance, a number of efforts have been made by the United Nations Environment Programme to build the capacity of domestic judges in terms of applying environmental principles at the grassroots. Such initiatives are aimed at the application of environmental principles at the domestic rather than the international level; see for instance, United Nations Environment Programme, *Judges Programme*, www.unep.org/Law/Programme_work/Judges_programme/index.asp; Environment News Service, *Judges Fortify Environmental Law Principles* (2002) www.ens-newswire.com/ens/aug2002/2002-08-27-01.asp.

[25] See the official website of the Global Compact, www.unglobalcompact.org.

[26] The Organisation for Economic Cooperation and Development, *The OECD Guidelines for Multinational Enterprises: Text, Commentary and Clarifications*, OECD Doc.DAFFE/IME/WPG(2000)15/FINAL (31 October 2001) ('*OECD Guidelines for Multinational Enterprises*').

[27] See Part I, art V, *OECD Guidelines for Multinational Enterprises*, above n 26.

[28] The *OECD Guidelines for Multinational Enterprises,* above n 26, at 29, point out that the provisions also take into account:

> 'the (Aarhus) Convention on Access to Information, Public Participation in Decision-making, and Access to Justice in Environmental Matters and reflects standards contained in such instruments as the ISO Standard on Environmental Management Systems.'

[29] See United Nations Global Compact, *The Ten Principles* (2008) www.unglobalcompact.org/AboutTheGC/TheTenPrinciples/index.html. The three environmental principles are: Principle 7, Businesses should support a precautionary approach to environmental challenges; Principle 8, undertake initiatives to promote greater environmental responsibility; and Principle 9, encourage the development and diffusion of environmentally friendly technologies. In relation to the Global Compact see chapter 8 of this book.

The fact that the *Stockholm Declaration* and the *Rio Declaration* list 26 and 27 environmental principles respectively does not necessarily mean that they all function in similar ways. The same environmental principles can receive different treatment from actors depending on the context in which they are employed. The precautionary principle, for instance appears in international agreements or treaties that are in force as part of international law.[30] However, the principle also gets frequent mention in instruments that do not, according to Article 38(1)(c) of the *Statute of the International Court of Justice,* reflect international law.[31] Some see the precautionary principle as an action-oriented rule,[32] a directing principle,[33] an environmental principle,[34] or an approach.[35] Yet there is doubt whether the precautionary principle is part of the customary practices of states and a norm of international law.[36] Whether it is a norm of international law or not does not

---

[30] Over 50 multilateral agreements have incorporated the precautionary principle in them: see Arie Trouwborst, *Evolution and Status of the Precautionary Principle in International Law* (2002) 63 *et seq*. Although the precautionary principle is used here other environmental principles are also deployed in similar ways in international law and politics. For instance the principle of common but differentiated responsibilities is the subject of very different use by scholars, diplomats and formally in international agreement; see Christopher Stone, 'Common but Differentiated Responsibilities in International Law' (2004) 98 *American Journal of International Law* 276.

[31] There are many such examples, but the most prominent of these is Principle 15 of the *Rio Declaration*. Art 38(1)(c) of the *Statute of the International Court of Justice* is relied upon as the authoritative statement regarding the sources of international law: see Ian Brownlie, *Principles of Public International Law* (5th ed, 2003) 3–4.

[32] Ulrich Beyerlin, 'Different Types of Norms in International Law' in Daniel Bodansky, Jutta Brunnée and Ellen Hey (eds), *The Oxford Handbook of International Environmental Law* (2007) 425, 440.

[33] de Sadeleer, above n 1, 266–274.

[34] Sands, above n 4, 231.

[35] Jacqueline Peel, 'Precaution – A Matter of Principle, Approach or Process?' (2004) 5 *Melbourne Journal of International Law* 483; Ellen Hey, 'The Precautionary Concept in Environmental Policy and Law: Institutionalizing Caution' (1991–1992) 4 *Georgetown International Environmental Law Review* 303; Birnie and Boyle, above n 5, 116.

[36] On this see generally Daniel Bodansky, 'Customary (and Not So Customary) International Environmental Law' (1995) 3 *Indiana Journal of Global Legal Studies* 105; Owen McIntyre and Thomas Mosedale, 'The Precautionary Principle as a Norm of Customary International Law' (1997) 9 *Journal of Environmental Law* 221. For arguments favouring its adoption as a norm of customary international law see, for instance, James Cameron and Juli Abouchar, 'The Status of the Precautionary Principle in International Law' in David Freestone and Ellen Hey (eds), *The Precautionary Principle and International Law: the Challenge*

appear to have an impact on the frequency of its use at the international level in terms of international politics.[37]

There are a range of environmental principles contained within treaties or various statements of international conferences and international organisations. Since the *Stockholm Declaration* and the *Rio Declaration* no other significant multilateral soft law instrument has sought to categorise the relevant principles of global environmental governance. Some environmental principles have received greater attention in international law and politics and have been the subject of increasing commentary and critique. Different scholars have developed their own lists of important environmental principles using different criteria. Sands has developed a list of environmental principles that he claims are significant because they are 'potentially applicable to all members of the international community across the range of activities which they carry out or authorise and in respect of the protection of all aspects of the environment'.[38] They include:

1. the obligation reflected in Principle 21 of the Stockholm Declaration and Principle 2 of the Rio Declaration, namely, that states have sovereignty over their natural resources and the responsibility not to cause transboundary environmental damage;
2. the principle of preventive action;
3. the principle of co-operation;
4. the principle of sustainable development;
5. the precautionary principle;
6. the polluter-pays principle; and
7. the principle of common but differentiated responsibility.[39]

---

of Implementation (1995) 29; Harald Hohmann, *Precautionary Legal Duties and Principles of Modern International Environmental Law: The Precautionary Principle* (1994) 12; Arie Trouwborst, 'The Precautionary Principle in General International law: Combating the Babylonian Confusion' (2007) 16(2) *Review of European Community and International Environmental Law* 185.

[37] This argument has been made in many different ways: see, for instance, Douglas Johnston, 'The challenge of international ocean governance: institutional, ethical and conceptual dilemmas' in Donald Rothwell and David VanderZwaag (eds), *Towards Principled Oceans Governance: Australian and Canadian approaches and challenges* (2006) 349; Stephen Toope, 'Formality and Informality' in Daniel Bodansky, Jutta Brunnée and Ellen Hey (eds), *The Oxford Handbook of International Environmental Law* (2007) 107.

[38] Sands, above n 4, 231.

[39] Ibid. It is important to point out that these particular principles are listed for illustrative purposes. Chapter 3 below discusses in more detail what environmental principles are and the constructed nature of the concept in international law and politics. Also, each principle listed here by Sands can appear in many different formulations across different multilateral agreements. For instance,

On the other hand Beder in a 2006 study examined six principles which had relevance for environmental policy makers.[40] In this list she included, aside from the sustainability, precautionary and polluter pays principles, the principles of participation and equity as well as human rights principles more generally.[41] What qualifies a particular or a group of norms as environmental principles appears to be conditional on the community of actors engaging with each other and the reason why someone is looking at a particular set of conditions.

The 1992 *Convention on Biological Diversity* states that:

> States have, in accordance with the Charter of the United Nations and the principles of international law, the sovereign right to exploit their own resources pursuant to their own environmental policies, and the responsibility to ensure that activities within their jurisdiction or control do not cause damage to the environment of other States or of areas beyond the limits of national jurisdiction.[42]

The *Biodiversity Convention* adopts only these two norms as the relevant environmental principles for that regime. Although it would be useful to develop a list of important environmental principles, each convention or treaty selects its own norms which are then classified as important and

David VanderZwaag has shown how the precautionary principle for instance appears in 14 different ways in multilateral agreements: see David VanderZwaag, 'The Precautionary Principle in Environmental Law and Policy: Elusive Rhetoric and First Embraces' (1999) 8 *Journal of Environmental Law and Practice* 355. Other lists developed within international institutional contexts have normative assumptions built into them. For instance, a report completed in 1996 by a group of experts within the United Nations Environment Programme in 1996 developed and argued for a list of environmental principles and concepts in international environmental law: Final Report of the Expert Group Workshop on International Environmental Law Aiming at Sustainable Development, Washington, DC, 30 September – 4 October 1996, UN Doc. UNEP/IEL/WS/3/2 (4 October 1996) Annex I. However the work of this group was geared to listing principles that are 'core elements' of the further development of international environmental law rather than existing norms that were being systematically applied by state actors: at [29]; see also, Beyerlin, above n 32, 429. See also the interesting efforts of Winfried Lang in developing a list of the more significant environmental principles: Winfried Lang, 'UN-Principles and International Environmental Law' (1999) 3 *Max Planck United Nations Yearbook* 157.

[40] Sharon Beder, *Environmental Principles and Polities: An Interdisciplinary Approach* (2006).

[41] Ibid, 2–6.

[42] Opened for signature 5 June 1992, 1760 UNTS 143 (entered into force 29 December 1993) ('*Biodiversity Convention*').

guiding principles for the regime. It is very possible for Sands' list of environmental principles to be used in a significant number of global environmental governance initiatives. As this book will go on to highlight and explore, particularly in the last chapter, it is more important to understand the nature of the way that environmental principles, as abstract and open-textured norms, influence governance through changing international law and politics. This is because a list of environmental principles will prove useful only if it is possible to appreciate how each of them influences governance and change more generally.

1.2 ENVIRONMENTAL PRINCIPLES, LEARNING AND CHANGE IN INTERNATIONAL LAW AND POLITICS

A lot has been written on the subject of principles, whether approached from an ethical, moral, legal, or political science perspective.[43] In scholarly works, the concept of a principle is often referred to as a norm that is 'imprecise' in terms of what it means,[44] or as having a wide margin of appreciation.[45] Braithwaite and Drahos have argued that principles as a type of norm prescribe 'highly unspecific actions'.[46] Miers and Twining have argued that definitions of principles are commonly united by some conception of the 'levels of generality, precision and prescriptive status of the norm'.[47]

Environmental principles generally share these characteristics with principles because often concepts within them are expressed in abstract

---

[43] As an example see the following special issues: (2002) 13(4) *European Journal of International Law*; (1996–1997) 82 *Iowa Law Review*; (1972) 81 *Yale Law Review* 799.

[44] Alexandre Kiss and Dinah Shelton, *Guide to International Environmental Law* (2007) 89.

[45] de Sadeleer, above n 1, 308.

[46] Braithwaite and Drahos, above n 2, 18.

[47] See William Twining and David Miers, *How To Do Things With Rules* (4th ed, 1999) 126. For instance, at 126, they write:

> From time to time, it may be useful to differentiate between general and specific rules, between vague and precise rules, between categorical precepts and mere guides or other standards which do not dictate results. Such distinctions have a bearing on problems of interpretation, but to insist on them at the start would introduce an artificial and premature rigidity into the discussion. Levels of generality, precision and prescriptive force are all matters of degree.

terms, or they are constructed as norms in an open-textured way.[48] After studying the principle of common heritage of mankind, Pinto argued that it took a '**quarter of century** . . . to determine by near-consensus the legal content of the "common heritage", only to have that legal content radically altered in the **two years** that followed'.[49] This suggests that there is a need to reconcile the abstract and open-textured nature of environmental principles with the fact that they have been, and are, commonly used in international law and politics. The dissonance between their abstract and open-textured qualities and their common use at the international level needs to be examined in terms of their potential to contribute to change in international law and politics.[50] This seems to have also been the message behind the following statement of the arbitral tribunal in the 2005 *Case concerning the Iron Rhine ('Ijzeren Rijn') Railway (Belgium v Netherlands)* where it was noted that:

> [t]here is considerable debate as to what, within the field of environmental law, constitutes 'rules' or 'principles'; what is 'soft law'; and which environmental treaty law or principles have contributed to the development of customary international law . . . The mere invocation of such matters does not, of course, provide the answer. . .[51]

---

[48] See chapter 3 of this book for a discussion of these terms.

[49] Moragodage Pinto, '"Common Heritage of Mankind": From Metaphor to Myth, and the Consequences of Constructive Ambiguity' in Jerzy Makarczyk (ed), *Theory of International Law at the Threshold of the 21st Century: Essays in Honour of Krzysztof Skubiszewski* (1996) 249, 265 (emphasis in original). Interestingly this comment assumes that the norm failed by assessing its effectiveness in terms of whether a uniform approach to what it means can or needs to be developed and agreed upon in international politics. See also Bodansky, 'Customary (and Not So Customary) International Environmental Law', above n 36, 115–116. Bodansky at 116 refers to environmental principles as representing 'the collective ideals of the international community, which at present have the quality of fictions or half-truths'. On the concept of common heritage of mankind and its application as an environmental principle see, for instance, Nico Schrijver, 'Permanent sovereignty over national resources versus the common heritage of mankind: complementary or contradictory principles of international economic law?' in Paul De Waart, Paul Peters and Erik Denters (eds), *International Law and Development* (1988) 87, 95–99.

[50] Beyerlin, above n 32, at 429 has also argued, after studying the views of certain scholars who have written on environmental principles, that 'the determination of the status, role, and effects of such "twilight" norms still needs clarification'.

[51] Arbitral Tribunal, Award of 24 May 2005, at paras 58 and 60, available through the web-page of the Permanent Court of Arbitration www.pca-cpa.org/upload/files/BE-NL%20Award%20240505.pdf.

The abstract and open-textured nature of environmental principles has led some scholars to classify environmental principles as indeterminate in terms of legal consequence,[52] as a 'twilight' norm at the 'bottom of the normative hierarchy of modern international environmental law',[53] or as 'general in the sense that they are potentially applicable to all members of the international community across the range of activities which they carry out or authorise and in respect of the protection of all aspects of the environment'.[54] Dhondt, writing about environmental principles in the context of the European Union, makes the observations that 'they have no direct legal consequences, require no specific action or they allow for derogation and are therefore not legally binding rules'.[55] These general views are sometimes even more pronounced when particular environmental principles, such as sustainable development, are examined more closely.[56]

As engaging as these critiques are they are mostly expressed in the context of whether environmental principles function in general terms as a legal rule or have legal consequences for actors.[57] This is problematic because of the assumption built into such views that normativity or the

---

52 See, for instance, de Sadeleer, above n 1, especially ch 4; Kiss and Shelton, above n 44, 89.

53 Beyerlin, above n 32, 426.

54 Sands, above n 4, 231.

55 Nele Dhondt, 'Environmental Law Principles and the Case Law of the Court of Justice' in Maurice Sheridan and Luc Lavrysen (eds), *Environmental Law Principles in Practice* (2002) 141, 153 (citations omitted, emphasis in original).

56 Different scholars have all picked up on the idea that sustainable development is general, vague or ambiguous in terms of what it means. See for instance, Magraw and Hawke, above n 10, 618–622; Gerhard Hafner, 'General Principles of Sustainable Development: From Soft Law to Hard Law' in Malgosia Fitzmaurice and Milena Szuniewicz (eds), *Exploitation of Natural Resources in the 21st Century* (2003) 53; Sumudu Atapattu, 'Sustainable Development, Myth or Reality? A Survey of Sustainable Development under International Law and Sri Lankan Law' (2001) 14 *Georgetown International Environmental Law Review* 265; Michael McCloskey, 'The Emperor Has No Clothes: The Conundrum of Sustainable Development' (1999) 9 *Duke Environmental Law and Policy Forum* 153; Colin Williams and Andrew Millington, 'The Diverse and Contested Meanings of Sustainable Development' (2004) 170(2) *The Geographical Journal* 99.

57 See, for instance, Alhaji Marong, 'From Rio to Johannesburg: Reflections on the Role of International Legal Norms in Sustainable Development' (2003) 16(1) *Georgetown International Environmental Law Review* 21, 44; Daniel Bodansky, 'Rules vs Standards in International Environmental Law' (2004) 98 *American Society of International Law Proceedings* 275; Beyerlin, above n 32; André Nollkaemper, 'Three Conceptions of the Integration Principle in International Environmental Law' in Andrea Lenschow (ed), *Environmental Policy Integration: Greening Sectoral Policies in Europe* (2000) 22.

consequences of norms are one-dimensional. Durkheim, for instance, in his study of the law, is said to have placed too much emphasis on its function in limiting the activities of individuals.[58] As a result he ignored the 'facilitative aspects of law', that is, 'the law concerned with powers, constituting relationships' and 'defining practices'.[59] The law can have an educational quality to it, and this is something that seems to be missing from Ehrlich's sociologically informed jurisprudence.[60] Endicott in his study of law as rules argued that they are inherently vague. However, because his study was contextualised in adjudicative approaches to the concept of law, its contribution to understanding norms is naturally limited.[61]

Koskenniemi has argued that environmental principles are often drafted with incompatible concepts and ideas built into them and actors are left to themselves to determine their meaning and normative impact at some future time.[62] This argument presumes that the critique about environmental principles is too abstract to draw conclusions about their normative potential. Assumedly the difficulty Koskenniemi is highlighting, presuming that he is correct that all environmental norms contain similar tensions within them, is in terms of the degree of uncertainty that comes with what environmental principles actually mean in practice given that they are open-textured. However, this does not say much about whether in contexts where certain conditions exist, such as environmental harm being very apparent, actors draw on environmental principles to learn how to develop their collective preferences. In other words, the open-textured nature of environmental principles might not contain the precise

---

58 Michael D A Freeman, *Lloyd's Introduction to Jurisprudence* (6th ed, 1994) 521.

59 Ibid.

60 Eugen Ehrlich, *Fundamental Principles of the Sociology of Law* (1936).

61 Timothy Endicott, *Vagueness in Law* (2000).

62 Koskenniemi, above n 16, 76. He gives the example of Principle 21 in the *Stockholm Declaration*, which gives states:

> the sovereign right to exploit their own resources pursuant to their own environmental policies, and the responsibility to ensure that activities within their jurisdiction or control do not cause damage to the environment of other States or of areas beyond the limits of national jurisdiction.

Koskenniemi points out that the principle states the values that underpin what states should do but also confirms their sovereignty. He has expressed similar views in his more recent scholarly works: see Martti Koskenniemi, 'The Fate of Public international Law: Between Technique and Politics' (2007) 70 *Modern Law Review* 1; Martti Koskenniemi, 'International Legislation Today: Limits and Possibilities' (2005) 23 *Wisconsin International Law Journal* 61.

normative implications for behaviour, but this does not stop groups from creating meaningful experiences from them in particular contexts.

The challenge of trying meaningfully to define a principle and the role it plays has meant that scholars have been developing typologies that distinguish them from other kinds of norms. This is in addition to separating different kinds of principles from each other.[63] De Sadeleer, for instance, differentiates environmental principles based on how they function. He writes that some principles 'indicate the essential characteristics of legal institutions (descriptive principles)', they can 'designate fundamental legal norms (basic principles)', or be viewed as filling 'gaps in positive law by assigning a constitutional or legal value to rules which are not yet formally set forth in written sources of law although they are considered essential (general principles of law)'.[64] Ebbesson on the other hand tries to avoid the dichotomy between environmental principles and rules so common in the scholarly work on legal norms by differentiating norms on the basis that they are either 'balancing norms', 'goal oriented norms' or 'fixed norms'.[65]

However, typologies in themselves may not adequately explain the potential role or function of environmental principles in terms of contributing to change in international law and politics. This can be seen in Beyerlin's analysis of environmental principles, where he seeks to distinguish them simply in terms of whether they are policies, legal principles, or legal rules.[66] He argues for instance that the precautionary principle is essentially an 'action-oriented rule'.[67] Sands argues that the core of the precautionary principle is 'still evolving' but maintains that it is an environmental principle.[68] Beyerlin and Sands actually adopt a similar definition of a principle in their work but come to different conclusions in relation to its application to the precautionary principle: for Sands the precautionary principle functions as an evaluative norm, whereas it is

---

[63] Typologies emerge from functionally differentiating between principles. For a typology of legal principles see Aleksander Peczenik, 'Principles of Law: The Search for Legal Theory' (1971) 2 *Rechtstheorie* 17; Manuel Atienza and Juan Ruiz Manero, *A Theory of Legal Sentences* (1998), especially chs 1 and 2.

[64] de Sadeleer, above n 1, 306; Kiss and Shelton, above n 44, 89.

[65] Jonas Ebbesson, *Compatibility of International and National Environmental Law* (1996) 83. As to the legal scholarly work on the dichotomy between rules and principles see Ronald Dworkin, *Taking Rights Seriously* (1977); Neil MacCormick, *Legal Reasoning and Legal Theory* (1978); Joseph Raz, *Practical Reason and Norms* (1999).

[66] Beyerlin, above n 32.

[67] Ibid, 440.

[68] Sands, above n 4, 268.

an 'action-oriented rule' for Beyerlin.[69] This suggests that presumptions about typologies explaining the role and function of environmental principles in different contexts are difficult to sustain because they rely on their interplay with social processes, like persuasion for instance, to contribute to change in international law and politics.

Even in situations where the role and function of environmental principles are examined, scholarship often does not distinguish between the consequences they could have for individuals as compared to groups of actors interacting at international level. Much attention has been given to environmental principles directing actors, obligating them to pursue certain ends, or changing their preferences.[70] The potential of environmental principles to change how groups of actors respond to environmental issues has not as a result been adequately explored.[71] The fact that this is the case seems odd, given that environmental principles have been frequently used in framework conventions which presume that ratifying states will further interact with each other to create meaning from them. Others have highlighted how environmental principles provide a sound basis for the development of international law,[72] but scholarship has not closely examined whether and how this is the case when it comes to groups of actors. In other words, descriptions of the potential role of environmental principles in framing and structuring what actors learn to be their preferred way of collectively responding to environmental issues need to be explored more rigorously.

The question examined in this book is whether and how, as an abstract

---

69 They both adopt what Ronald Dworkin has written about a principle. That is, it is 'one which officials must take into account, if it is relevant, as a consideration inclining in one direction or another': Dworkin, above n 65, 26. It is a norm used to evaluate the conduct of actors: Bodansky, above n 57, 277.

70 Christian Reus-Smit, 'Introduction' in Christian Reus-Smit (ed), *The Politics of International Law* (2004) 1, 3.

71 This is not to suggest that scholarship in relation to principles does not focus on how communities develop preferences, but that it is more focused on areas of the world that actors share in common with each other. Brunnée as an example discusses the concepts of 'common areas', 'common Heritage' and 'common concern' which have shaped how states collectively respond to environmental concerns: Jutta Brunnée, 'Common Areas, Common Heritage, and Common Concern' in Daniel Bodansky, Jutta Brunnée and Ellen Hey (eds), *The Oxford Handbook of International Environmental Law* (2007) 550.

72 See the statement by Hans Alders at the United Nations Conference on the Environment and Development in 1992 during which he was the Minister of Environment of the Netherlands: 'We see the Declaration as a sound basis for the much needed development of international law, and therefore endorse it as its stands', quoted in Panjabi, above n 7, 273.

and open-textured norm, environmental principles contribute to changing the rules and norms of international law and politics. The presumption that an environmental principle is a legal, value based or moral norm narrows the diverse ways that it can generate, facilitate, structure or frame change in the intersubjectivity amongst groups at the international level. It has been argued that international law is incapable of differentiating between the casual pathways that a norm requires to create change.[73] This suggests that the approach to studying principles is as important as the questions whether they can contribute to change or not.

1.3 GENERAL OUTLINE AND STRUCTURE OF THE BOOK

The approach taken to the questions just posed can be broadly characterised as falling within the field of scholarship pertaining to the regulation of what transnational actors do, as opposed to public international law which focuses more narrowly on states and their cooperation problems.[74] Scholarship relating to how transnational actors develop interests and preferences in relation to the environment and how they can best be influenced or determined is gaining momentum.[75] The potential range of

---

[73] This roughly translates the difference in approach between liberal and social constructivists to the politics of international law: see Christian Reus-Smit, 'The politics of international law' in Christian Reus-Smit (ed), *The Politics of International Law* (2004) 14, 15–24.

[74] The term regulation can be misleading a little because this book focuses on both the regulatory and constitutive role of environmental principles; in relation to the distinction between regulatory and constitutive effects of norms see, for instance, Nina Tannenwald, 'The Nuclear Taboo: The United States and the Normative Basis of Nuclear Non-use' (1999) 53(3) *International Organization* 433. A difference in the approach to regulation can be seen by comparing the works of Hilary Charlesworth and Christine Chinkin, 'Regulatory Frameworks in International Law' in Christine Parker, Colin Scott, Nicola Lacey and John Braithwaite (eds), *Regulating Law* (2004) 246, and Braithwaite and Drahos, above n 2. Braithwaite and Drahos focus more broadly on norms, mechanisms for regulation and actors rather than states and international law rules.

[75] The term 'transnational actors' refers to states and other actors like transnational corporations. On regulation and the environment see, for example, Richard Stewart, 'Instrument Choice' in Daniel Bodansky, Jutta Brunnée and Ellen Hey (eds), *The Oxford Handbook of International Environmental Law* (2007) 147; Scott Barrett, *Environment and Statecraft: The Strategy of Environmental Treaty-Making* (2003); Jonathan Wiener, 'Global Environmental Regulation: Instrument Choice in Legal Context' (1999) 108 *Yale Law Journal* 677; Richard Stewart, 'A New Generation of Environmental Regulation?' (2001) 29 *Capital University*

conclusions that one can draw from case studies depends on the perspective taken in studying regulation. This means that structural differences between the international and domestic systems make it difficult and inappropriate to apply theories of regulation from the domestic level to transnational actors.[76]

It is also difficult to use certain approaches to regulation developed by scholars of international law because of the particular theoretical focus, from an international relations perspective, that they adopt in framing their study. For instance, Barrett argues that it would be difficult for environmental agreements to achieve any significant impact on states without appropriate carrots and/or sticks to support the use of this kind of a regulatory instrument.[77] This approach to regulation presumes that some kind of coercive compulsion or 'material lever' is necessary for a treaty to achieve significant results.[78] Alternative theoretical approaches to international relations, such as social constructivism, assume that social processes, like 'authentic persuasion', can help actors as a group or a community to build intersubjectivity to better define what they each want.[79] Actors learn about what they are interested in by participating in and engaging with others in social contexts established at the international level.

This book adopts and develops the moderate social constructivist approach to international relations and politics which theoretically shapes how certain issues are dealt with.[80] Much of the sociologically informed

---

Law Review 21; Nathaniel Keohane, Richard Revesz and Robert Stavins, 'The Choice of Regulatory Instruments in Environmental Policy' (1998) 22 *Harvard Environmental Law Review* 313; Organisation for Economic Cooperation and Development, *Handbook of Incentive Measures for Biodiversity: Design and Implementation* (1999); Philippe Sands, *Transnational Environmental Law: Lessons in Global Change* (1999).

76 Richard Stewart, 'Instrument Choice' in Bodansky, Daniel, Brunnée, Jutla, and Hey, Ellen (eds), *The Oxford Handbook of International Environmental Law* (2007) 147 at 161–168.

77 Barrett, above n 75.

78 See Thomas Schelling, *Arms and Influence* (1966); referred to in Rodger Payne, 'Persuasion, Frames and Norm Construction' (2001) 7 *European Journal of International Relations* 37, 41.

79 Payne, ibid.

80 See, for instance, John Ruggie, *Constructing the World Polity: Essays on internationalization institutionalization* (1998); Emanuel Adler, *Communitarian International Relations: The Epistemic Foundations of International Relations* (2005); Reus-Smit, above n 73; Jeffrey Checkel, 'Why Comply? Social Learning and European Identity Change' (2001) 55(3) *International Organization* 553; Alexander Wendt, *Social Theory of International Politics* (1999). The constructivist approach adopted for this book is described in chapter 2.

scholarly work on norms at the international level has engaged with the differences between social and legal norms,[81] whether they matter,[82] and the social processes through which they can influence the behaviour of actors.[83] Although this book draws on these views to examine and critique existing approaches to the role and function of environmental principles it is also concerned with the qualitative dimensions of how they contribute to change at the international level. In this sense, this work examines how groups of actors learn in social contexts established at the international level.[84] The idea that groups of actors can learn what their preferences are through their interactions at the international level emerges from the social constructivist approaches developed from international relations and politics.[85] Processes such as persuasion and social influence are dimensions of the learning experiences of groups at the international level. Other dimensions include how norms like environmental principles, by being open-textured and abstract, interplay with social processes to help create the learning experiences of actors. The focus on collective learning and the social processes that help create them means that this book should be able to offer views other than whether environmental principles matter or not.

What this means is that this book is not concerned with a number of other things. It does not say much about what particular environmental principles mean because it will examine how they are expressed within the context of learning experiences that actors have at the international level. Although this book is about the role of environmental principles in global governance it does not seek to determine their source or how they originate because this work is more about the constructed nature of their function in international law and politics. In other words it is not concerned with the historical origins of environmental principles and whether the context of how they developed defines them as a legal, political or a moral norm.

---

81 See Reus-Smit, above n 70, 3. On this see also Peter Katzenstein (ed), *The Culture of National Security: Norms and Identity in World Politics* (1996); Martha Finnemore, 'Are Legal Norms Distinctive?' (2000) 32 *New York University Journal of International Law and Politics* 699; Friedrich Kratochwil, *Rules, Norms, and Decisions: On the Conditions of Practical and Legal Reasoning in International Relations and Domestic Affairs* (1991).

82 Martha Finnemore and Kathryn Sikkink, 'International norm dynamics and political change' (1998) 52(4) *International Organization* 887.

83 Ibid.

84 See chapter 2 of this book for an explanation of these ideas and concepts and how they link together for the purposes of this work.

85 In particular, see Payne, above n 78; Martha Finnemore, *National Interests in International Society* (1996), especially ch 1.

Importantly, this book does not differentiate between change in international law and politics. It uses Reus-Smit's ideas, which emerge out of his social constructivist approach to social reality, on the mutually 'constitutive relationship between international politics and law'.[86] He refers to politics as a 'variegated, multi-dimensional form of human deliberation and action'.[87] In this sense, he argues that politics integrates four types of reasoning: idiographic, purposive, ethical, and instrumental.[88] When actors deal with questions like 'how should we act?', they are engaging in politics with a particular purpose in mind.[89] As will be discussed in this work it is through the interplay of social processes like persuasion with the abstract and open-textured qualities of norms like environmental principles that meaning is created in context. This suggests that developing a strict dichotomy between international law and politics can serve no real purpose when one is studying the function and role of environmental principles at the international level. To separate law and politics would actually restrict or narrow the kind of change that could be observed at the international level. For instance, intersubjectivity that emerges from actors engaging with each other might take the form of instantiating a cultural norm describing what international law and politics are capable of achieving.[90]

Lastly, this book uses the term 'actor' to avoid having to be specific about whether it refers to a state, transnational corporation or non-governmental organisation. This is because the focus of this work is on examining the role and function of environmental principles in bringing about change in terms of whether groups learn at the international level. The work does not take for granted that it is arguable whether transnational corporations are capable of political association at the international level.[91] It is important, however, for this work to highlight the potential that environmental principles have for influencing transnational

---

[86] Reus-Smit, above n 73, 14. See also Christian Reus-Smit, 'Politics and International Legal Obligation' (2003) 9 *European Journal of International Relations* 591, and Alexander Wendt, who argues that we need to look at how politics is socially constructed or constituted rather than assume that it means only one thing in international relations: Alexander Wendt, *Social Theory of International Politics* (1999), especially ch 1.

[87] Reus-Smit, above n 73, 24.

[88] Ibid, 25.

[89] Ibid.

[90] For an example of this argument see section 7.3. and chapter 7 more generally.

[91] For this see section 7.3. and chapter 7 more generally.

corporations which are significant actors at the international level and in shaping politics.[92]

1.3.1 The Structure of the Book

This book is divided into four parts. The first, which includes this introduction, discusses the theoretical framework which has given shape to this work. Chapter 2, after briefly discussing moderate social constructivism, examines how the idea of social learning is different from socialisation. In particular it highlights how the practices of actors give meaning to social learning. It also establishes the importance of steering social learning and briefly discusses how persuasion and social influence are important processes in this respect.

The second part assesses what the concept of an environmental principle is and how different scholarly works have approached ideas relating to its role and function. Chapter 3 defines environmental principles and explores what it means for them to be abstract and open-textured. It uses the concept of social learning amongst groups to assess what they are rather than whether environmental principles are legal, moral, ethical, or political norms. This distinguishes this study of environmental principles from other works on norms within international law and relations. Chapter 4 discusses a selection of scholarly approaches to social learning which have valued the abstract and open-textured nature of norms more generally. It seeks to highlight their differences and similarities to the approach adopted in this work to examining principles in the context of social learning. Chapter 5 reviews the scholarship on the function and role of environmental principles at the international level to highlight some significant realist and liberal approaches to the ways that environmental principles contribute to change in international law and politics. This establishes the significance of the definition of an environmental principle adopted in chapter 3 by highlighting the potential it has for understanding their role in changing international law and politics.

Part III studies the role and function of environmental principles within particular groups of actors, drawing on social processes that have a dominant role within their interactions. The selected case studies are different from each other in many ways because this part of the book aims to

---

92 The lack of attention given to transnational corporations in the development of law and politics internationally is an important issue: see Fleur John, 'The Invisibility of the Transnational Corporation: An Analysis of International Law and Legal Theory' (1994) 19 *Melbourne University Law Review* 893.

examine whether environmental principles play a role in changing international law and politics rather than whether they do so in one particular context. Also, given the variable nature of how groups learn, the selected chapters each discuss a different social process that is either persuasive or influences how groups intersubjectively identify with each other. The purpose of this is not only to examine whether environmental principles interplay with or constitute a variety of processes, but also to identify the variable role they play in this respect in different contexts. Additionally, this part also adopts case studies that identify different types and kinds of actors interacting with each other at the international level. The aim here, once again, is to ensure that the arguments in this book are not developed in relation only to one type of actor, given that international law and politics engage more than just states.

Chapter 6 examines the diffuse nature of power as a persuasive process within the interactions of the International Court of Justice (ICJ). It selects a variety of ways in which environmental principles arguably interplay with or constitute power within and through the ICJ. Given that individuals are judges on the ICJ and only states can litigate before it, Chapter 7 focuses on the Global Compact (GC) initiative through the United Nations (UN) which has corporations as its participants. It examines how diffusion as a social process interplays with environmental principles to instantiate a culture shift through the GC. It also assesses the potential that environmental principles have to constitute the type of identity that membership of the GC might be generating. Chapter 8 studies the role and function of environmental principles in the negotiations of the contracting parties to the 1972 *Convention on the Prevention of Marine Pollution by Dumping of Wastes and Other Materials*,[93] and the 1996 *Protocol to the Convention on the Prevention of Marine Pollution by Dumping of Wastes and Other Matter*,[94] relating to the injection of carbon dioxide (CO_2) into the sub-seabed repositories. It examines the interplay between the precautionary principles and the persuasive nature of arguments as opposed to bargaining as a style of negotiation.

---

[93] Opened for signature 29 December 1972, 1046 UNTS 120 (entered into force 30 August 1975) ('LC'). The agreement was renamed in 1993 the *London Convention.*

[94] Opened for signature 7 November 1996, (1997) 36 ILM 1 (entered into force 24 March 2006) ('LP'). The term ocean dumping regime will be used to refer to the collective impact of the LC, LP, and the resolutions of the consultative committee to both agreements. The LP as of 31 March 2008 has 34 contracting parties compared to the 82 that have signed and ratified the LC. See the official website of the International Maritime Organisation at www.imo.org/

The last part is chapter 9 which concludes this work. It discusses the importance of social learning as the dynamic for change in international law and politics. In particular it draws from the three case studies to show how the interplay between environmental principles and processes of persuasion or social influence can generate common knowledge or cultures amongst groups of actors as a result of their interactions with each other. Importantly the chapter discusses the significance of environmental principles through their role and function in social learning. It concludes by identifying the potential importance of environmental principles in protecting the environment at the international level.

2. Social learning at the international level

2.1 INTRODUCTION

The role of norms at the international level can be studied in a variety of ways, ranging from realist, liberal, social constructivist and post-modern approaches to ontology and epistemology. This is also complicated by the fact that there are also varying philosophical approaches to the controversies relating to ecological issues and the environment more generally.[1] This rather complicated landscape of normative and philosophical differences for international relations and environmental ethics has an impact on how one studies the role and function of environmental principles.

This chapter sets out in general terms the social constructivist approach that frames the discussion of social learning in this work.[2] Its aim is to discuss the idea of collective or communitarian social learning in international politics and to identify how we can recognise it in a meaningful way. The terms common and collective knowledge are used in this chapter and work, using the scholarship of Wendt,[3] to describe the micro and macro

---

[1] Christopher Stone, 'The Environment in Moral Thought' (1988) 56 *Tennessee Law Review* 1; Christopher Stone, '*Should Trees Have Standing?* Revisited: How Far Will Law and Morals Reach? A Pluralist Perspective' (1985) 59 *Southern California Law Review* 1; Joseph DesJardins, *Environmental Ethics: An Introduction to Environmental Philosophy* (4th ed, 2005).

[2] It should be noted that the version of social constructivism used in this book has strong roots and correlations with what some have called mainstream American constructivism: see Ole Wæver, 'Four Meanings of International Society: A Trans-Atlantic Dialogue' in Barbara Allen Roberson (ed), *International Society and the Development of International Relations Theory* (1998) 80.

[3] Alexander Wendt, *Social Theory of International Politics* (1999). The work of Wendt has been influential in international relations particularly in developing the moderate version which is used in this work. For other scholarly works of Wendt see: Alexander Wendt, 'Driving with the Rearview Mirror: On the Rational Science of Institutional Design' (2001) 55(4) *International Organization* 1019; Alexander Wendt, 'Collective Identity Formation and the International State' (1994) 88(2) *American Political Science Review* 384; Alexander Wendt, 'The Agent-structure Problem in International Relations Theory' (1987) 41 *International Organization*

potential of social learning. That is, we can recognise social learning in terms of how common knowledge, which may take the form of a rule, is expressed within particular contexts.[4]

Importantly, this chapter argues that communitarian approaches to social learning depend on the nature or quality of the participation or the social practices of actors. The next chapter discusses how environmental principles have the capacity socially to frame or structure these practices. This chapter, however, examines how the nature of the participation of actors has the potential to steer collective social learning. It is detailed in its explanation of the various ideas and the way that they are linked because the concept of collective social learning is still emerging as a field of scholarship within social constructivism in international relations. This chapter conceptually develops the idea of collective social learning in the context of social practice related to change in international law and politics. This is important because conceptions of abstract and open-textured norms in constructivist works have limitations when studied within a framework of learning.

335; Alexander Wendt, 'Anarchy is what States Make of it: The Social Construction of Power Politics' (1992) 46 *International Organization* 391; Alexander Wendt, 'Constructing International Politics' (1995) 20 *International Security* 71. For favourable and critical engagements with Wendt's work see: Stefano Guzzini and Anna Leander (eds), *Constructivism and International Relations. Alexander Wendt and His Critics* (2005); and also the entire volume of (2000) 26 *Review of International Studies* 123–180 focused on Wendt's Social *Theory of International Politics* (1999).

4 Common knowledge is 'shared mental models': Arthur Denzau and Douglass North, 'Shared Mental Models: Ideologies and Institutions' (1994) 47 *Kyklos* 3; Wendt, above n 3, 161. These can take a variety of different forms when they are expressed or reified. John Dryzek, for instance, when talking about discourse as intersubjectivity, states the following as expressions of it:

> Discourses establish meanings, identify agents in contrast to those who can only be the object of action, confirm relations between actors and other entities, set the boundaries for what is legitimate knowledge, and generate what is accepted as common sense.

John Dryzek, 'Paradigms and Discourses' in Daniel Bodansky, Jutta Brunnée and Ellen Hey (eds), *The Oxford Handbook of International Environmental Law* (2007) 44, 46; citing Jennifer Milliken, 'The Study of Discourse in International Relations: A Critique of Research and Methods' (1999) 5 *European Journal of International Relations* 225.

2.2 SOCIAL CONSTRUCTIVISM AND INTERNATIONAL RELATIONS

In general terms, this work adopts what Checkel classed as a 'modernist' view of social constructivism in international relations, which combines 'an ontological stance critical of methodological individualism with a loosely causal epistemology'.[5] There are different ways one can read the social constructivist literature and characterise its main conceptual tools or framework approaches to international relations. There are some who develop the concepts that drive social constructivism and others who espouse particular theories based on varying ontological and epistemological vantage points. Ruggie, for instance, characterises theoretical approaches to social constructivism into three categories: neo-classical, post-modernist, and naturalist constructivism.[6] According to him, neo-classical constructivist thought is concerned with identifying analytical tools that help make sense of intersubjective meanings that are developed in the social fabric of international life. Some of these analytical tools include speech act theory, and the theory of communication action.[7] Post-modern constructivists drawing from Foucault and Derrida offer a different insight into international relations by stressing the discursive practices that construct the subject.[8] It is through various discursive practices that the ontological primitive is constituted in a particular way in international society.[9]

Finally, naturalistic constructivism draws from both neo-classical and post-modern constructivism but is rooted in the scientific realism developed by Bhaskar starting in 1979.[10] Scientific realism assumes that

---

[5] Jeffrey Checkel, 'Why Comply? Social Learning and European Identity Change' (2001) 55(3) *International Organization* 553, 554.

[6] John Ruggie, *Constructing the World Polity: Essays on international institutionalization* (1998) 35–36.

[7] In addition to himself, some of the writers John Ruggie classes as neo-classical social constructivists are Ernst Haas, Fredrich Kratochwil, Nicholas Onuf, Emanuel Adler, Martha Finnemore, Peter Katezenstein, and feminist scholars like Jean Elshtain.

[8] Ruggie, above n 6, 35.

[9] Ibid. Ruggie classes the following scholars as post-modern constructivist: Richard Ashley, David Campbell, James Der Derian, R B J Walker and feminist scholars such as Spike Peterson. Christian Reus-Smit (*The Moral Purpose of the State* (1999)) can also be added to this list.

[10] See Roy Bhaskar, *Scientific Realism and Human Emancipation* (1986). In relation to scientific realism see also Margaret Archer, *Realist Social Theory: The Morphogenetic Approach* (1995).

intersubjective beliefs sometimes exist independently of the individuals whom it constitutes.[11] This belief flows from the observation that both the physical and ideological world contain non-observables, which at times exert themselves and help constitute intersubjective beliefs. The use of scientific realism goes against the idea that our reality is essentially a social construct, and Wendt as the dominant scholar in this genre accepts this. However the use of scientific realism finds a place for brute material forces which seems more plausible than some other constructivist approaches.

This work adopts an eclectic mix of approaches to constructivism developed into a prism to view social learning which is discussed in the sections below. It casts its approach to social constructivism against the background of neo-realist and neo-liberal approaches to international relations.[12] Constructivists, to put things simplistically, argue that international actors are not just 'rational actors in pursuit of efficient means to realise individual and collective interests'.[13] This means that communities or interactions amongst actors, as inherently social and intersubjective processes, have causal and constitutive effects on actors. Other works in the field of international relations place the value of research solely in trying to understand the nature of international politics by studying the rational choices of actors as often driven by internal constituencies and the make-up of domestic politics. Generally speaking, social constructivism focuses on ontology and epistemology in a different way from 'rationalist and relativist interpretive approaches' to international relations.[14]

The real similarity between social constructivists is in their emphasis on the ontological importance of the 'social world as intersubjectivity and collectively meaningful structures and processes'.[15] This means that what is said and done within particular environments will be meaningful to actors because the foundations of their communities are built on knowledge, ideas or epistemes. It is the practice of actors within their

---

[11] Ruggie, above n 6, 35–36.

[12] For discussions of the contrast between realism, rational institutionalism and social constructivism see: Wendt, above n 3; Ruggie, above n 6.

[13] David Wippman, 'The International Criminal Court' in Christian Reus-Smit (ed), *The Politics of International Law* (2004) 151, 152.

[14] Emanuel Adler, *Communitarian International Relations: The Epistemic Foundations of International Relations* (2005) 92. For instance, this is evident in works that approach social learning for states from a rational choice perspective: see Andrew Farkas, *State Learning and International Change* (2001).

[15] Adler, above n 14, 11.

intersubjectively determined communities that produces culture and acts constitutively on them.[16]

For social constructivists it is necessary to emphasise the contingent nature of reality and our generalisations of it.[17] It opens up the possibility that our engagement with the world can create new understandings of realities that are yet to become a part of our social reality. That is, even if material realities exist they may not have a cultural dimension which creates meaning for us.[18] It is our interactions with each other that create social meaning and structure the way we view the world (including material objects and phenomena) around us. The emphasis of social constructivism on culture means that the 'intersubjective knowledge and ideas' of communities are not only an important part of reality but can also have a variety of different effects. For instance they can 'have constitutive effects on social reality and its evolution'.[19] This next section describes the idea of common and collective culture, described by Wendt in his work on *Social Theory of International Politics* as two different examples of social structure or intersubjectively established ways of viewing the world.

2.2.1 Common and Collective Culture

To suggest that international relations is constituted by culture and ideas which are intersubjectively shared says little about the level at which it might operate. For instance, the English School of international relations has described the potential that shared norms have in bringing order to the world of states.[20] Scholars from within that school of thought operate at

---

[16] On the idea of communities of practice see: ibid; Etienne Wenger, *Communities of Practice: Learning, Meaning, and Identity* (1999).

[17] Adler, above n 14, 12.

[18] Wendt, above n 3, 158.

[19] Adler, above n 14, 12.

[20] Barry Buzan, *From International to World Society? English School Theory and the Social Structure of Globalisation* (2004), 7. For more on the English School, particularly in its relationship with social constructivism in international relations, see Timothy Dunne, 'International Society: Theoretical Promises Fulfilled?' (1995) 30 *Cooperation and Conflict* 125; Timothy Dunne, *Inventing International Society: A History of the English School* (1998); Timothy Dunne, 'New Thinking on International Society' (2001) 3(2) *British Journal of Politics and International Relations* 223; Andrew Hurrell, 'Norms and Ethics in International Relations' in Walter Carlsnaes, Thomas Risse-Kappen and Beth Simmons (eds), *Handbook of International Relations* (2005); Andrew Hurrell, 'Foreword to the Third Edition: The Anarchical Society 25 Years On' in Hedley Bull, *The Anarchical Society: a Study of Order in World Politics* (3rd ed, 2002); Kai Alderson and Andrew Hurrell (eds), *Hedley Bull on International Society* (2000).

the level of what they term 'international society'.[21] Wendt has developed a distinction between what he terms common and collective knowledge, as part of his constructivist approach to international politics, to describe two different kinds of structure.[22] These concepts reinforce the methodologically holistic view of how, according to some social constructivists, we should study reality.[23]

Common knowledge describes what actors believe about the 'rationality, strategies, preferences, and beliefs' of other international actors.[24] Wendt points out that these beliefs need not be true but only believed to be true.[25] In this context, then, these beliefs about what others think will matter when they 'interlock' with other actors' beliefs.[26] In this sense, Wendt points out that common knowledge is the same as having an 'intersubjective understanding' of a situation. Common knowledge is expressed in cultural forms like 'norms, rules, institutions', and 'ideologies'.[27] He has argued that common knowledge is constitutive of the actors' interests, but because it is a view held by the individual it is also transient in the impact it has on the individual.[28]

Collective culture on the other hand is supervenient or, citing Durkheim, he points to it as '"collective" representations or knowledge'.[29] Wendt argues that these 'knowledge structures' are 'held by groups which generate macro-level patterns in individual behaviour over time'. The connection of collective knowledge to individuals is that they must believe in it, and in that process they reproduce the idea. Collective knowledge, in other words, is still a belief and not a reality that is absolute for all time. However the effects of collective knowledge cannot be reduced to what individuals might believe. The example Wendt gives is capitalism or the Westphalian system.[30] These collectively held ideas are not reducible to one actor and survive over time. In the context of this work, the idea that we protect the environment for future generations is so taken for granted

---

21 Buzan has argued that they operate at three different levels: international system, international society and world society: see Buzan, above n 20, 7.

22 Wendt, above n 3, 158.

23 See eg, ibid, 26–29; Adler, above n 14, 12–13.

24 Wendt, above n 3, 159. It must be pointed out that Wendt uses this concept in relation to states.

25 Ibid, 160.

26 On the concept of interlocking beliefs see Rajeev Bhargava, *Individualism in Social Science* (1992).

27 Wendt, above n 3, 160.

28 Ibid, 161.

29 Ibid.

30 Ibid.

that it is arguably a collectively held idea that is now supervenient rather than reducible to what one actor might think internationally. That is, although we take for granted the idea that we have to protect the environment for the future generation, this is a philosophical position rather than an expression of the truth.[31]

This distinction between common knowledge and collective knowledge is varyingly described in philosophy, sociology, psychology and social sciences more generally. As an example, in relation to environmental law, Dryzek has argued that the terms discourse and paradigm can be used as descriptions of an 'inter-subjective understanding that conditions individual action, and social outcomes, in the international system'.[32] He describes a discourse as 'composed of shared concepts, categories, and ideas that enable actors to understand situations'.[33] A paradigm on the other hand ideologically 'structures the world and, in particular, its problems or aspects requiring explanation'.[34] It is difficult to argue that this distinction corresponds exactly with the concepts of common knowledge and collective culture because of the difference in vocabulary and disciplinary interests.[35]

However, whether the term discourse or common knowledge is used, what is relevant for this study is the variety of different ways that we can observe intersubjectivity and the kind of reification that is used to express it. Therefore a norm is either an expression of discourse or is common knowledge amongst actors because their beliefs about what is appropriate interlock or are shared intersubjectively. Secondly, it is also about the way that common knowledge or discourse affects the individual. That is, common knowledge or discourse can either have an effect on the behaviour of actors within particular situations or can constitute their identities and therefore who they see themselves as.[36] This in turn can affect how actors relate to each other and to the social environments around them.

---

31 See Joseph DesJardins *Environmental Ethics: An Introduction to Environmental Philosophy* (4th ed, 2005) 74–83.

32 Dryzek, above n 4, 44–62. See also John Dryzek, *The Politics of the Earth: Environmental Discourses* (2nd ed, 2005).

33 Dryzek, above n 4, 46. In international relations, see, for instance, Jennifer Milliken, 'The Study of Discourse in International Relations: A Critique of Research and Methods' (1999) 5 *European Journal of International Relations* 225.

34 Dryzek, above n 4, 45.

35 Dryzek, ibid, 46, has pointed out that an 'emphasis on the constitute role of discourses is consistent with a constructivist position in international relations theory (though constructivists do not have to be discourse analysts).'

36 See Nina Tannenwald 'The Nuclear Taboo: The United States and the Normative Basis of Nuclear Non-use' (1999) 53(3) *International Organization* 433,

Before concluding this general description of social constructivist thought it is important to state what the role and function or, as Wendt puts it, the effects of structure are. It is generally assumed that rational choice is best placed to explain the role of ideas in terms of individual actors.[37] Wendt argues that the difference between rational choice theory and constructivism has more to do with the manner in which agents' preferences and identities are constructed.[38] Social constructivists, he argues, focus on the constitutive effects of culture on identities and interests/preferences, whereas rational choice theorists focus on behaviour.[39] In this sense, constructivists argue that the interaction of actors within social environments constructs their preference and/or their identities as opposed to just causing them to behave in a particular way which may change with the times.[40] This discussion is used in the next few sections to develop a conception of social learning as the dynamic for change in international law and politics.

2.3 SOCIAL LEARNING AS A DYNAMIC FOR CHANGE

The use of social learning as a concept for studying social change in international relations and politics is not new. Although the study of the role and function of norms at the international level is also not new,[41] the use of social learning as the context for it is of increasing interest to scholars in social constructivism.[42] The notion that groups of actors learn, or that

---

for a discussion of the differences between the regulatory and constitutive effects of intersubjectively shared structures of thought such as norms.

37 Wendt, above n 3, 166.

38 Ibid.

39 Ibid, 168. On the idea of constitutive causality see also Alexander Wendt, 'On constitution and causation in International Relations' (1998) 24(5) *Review of International Studies* 101; Emanuel Adler, 'Seizing the Middle Ground: Constructivism in World Politics' (1997) 3(3) *European Journal of International Relations* 319.

40 On the causal effects of structure on agents see, for instance, Norman Fairclough, *Analysing Discourse: Textual Analysis for Social Discourse* (2003); Emanuel Adler and Steven Bernstein, 'Knowledge in Power: the Epistemic Construction of Global Governance' in Michael Barnett and Raymond Duvall (eds), *Power in Global Governance* (2005) 294, 296; Wendt, above n 3, 167–190.

41 Andrew Cortell and James Davis Jr, 'Understanding the Domestic Impact of International Norms: A Research Agenda' (2000) 2(1) *International Studies Review* 65.

42 See, for instance, Trine Flockhart, '"Complex Socialization": A Framework

their cognition collectively evolves or changes, has been of particular interest for social constructivists in international relations.[43] Social learning in this work is defined as the change of the common knowledge or the instantiation of collective beliefs that groups of actors develop intersubjectively. It is the development or the construction of intersubjectively shared knowledge and ideas or the instantiation by groups of an emerging culture.

The discussion below is therefore separated between first identifying what collective or communitarian social learning is and then focusing on processes that steer and give it direction. The focus on the collective nature of the participation of actors is necessary because of the constructivist

---

for the Study of State Socialization' (2006) 12(1) *European Journal of International Relations* 89; Kai Alderson, 'Making sense of state socialization' (2001) 27 *Review of International Studies* 415; Martha Finnemore and Kathryn Sikkink, 'International norm dynamics and political change' (1998) 52(4) *International Organization* 887; Martha Finnemore, 'International organizations as teachers of norms: the United Nations Educational, Scientific, and Cultural Organization and science policy' (1993) 47 *International Organization* 565; Checkel, above n 5; Jeffrey Checkel, 'International Institutions and Socialization in Europe: Introduction and Framework' (2005) 59(4) *International Organization* 801. In the context of international relations, discussions of social learning are now heavily contextualised within a variety of different conceptual fields. See for instance these various works discussing social learning: Daniel Beland, 'The Politics of Social Learning: Finance, Institutions, and Pension Reform in the United States and Canada' (2006) 19 *Governance* 559; Clark Miller, 'Democratization, International Knowledge Institutions, and Global Governance' (2007) 20 *Governance* 325; Thomas Risse, Stephen Ropp and Kathryn Sikkink (eds), *The Power of Human Rights: International Norms and Domestic Change* (1999). This list does not intentionally ignore the scholarship of the English School in international relations. The concept of international society in the English School is particularly well developed and focuses on the institutionalisation of shared norms.

43 See, for instance, Ronald Jepperson, Alexander Wendt, and Peter Katzenstein, 'Norms, Identity, and Culture in National Security' in Peter Katzenstein (ed), *The Culture of National Security. Norms and Identity in World Politics* (1996) 33; Adler and Bernstein, above n 40; Paul Edwards, 'Representing the Global Atmosphere: Computer Models, Data, and Knowledge about Climate Change' in Clark Miller and Paul Edwards (eds), *Changing the Atmosphere: Expert Knowledge and Environmental Governance* (2001) 31; Miller, above n 42, 325–357; Nina Tannenwald, 'The Nuclear Taboo: The United States and the Normative Basis of Nuclear Non-Use' (1999) 53(3) *International Organization* 433; Martha Finnemore, *National Interests in International Society* (1996). The idea that international society is bound by norms which bring states together in solidarity is a common idea to some scholars who write in the English School of international relations tradition: for example, Hedley Bull, *The Anarchical Society: A Study of Order in World Politics* (1977); Hedley Bull, 'Natural Law and International Relations' (1970) 5(2) *British Journal of International Studies* 171; Buzan, above n 20.

approach to learning which places emphasis on how mental structures of thought are constructed intersubjectively as different actors engage with the environment around them.[44] As will be argued elsewhere,[45] concepts such as consent, obligation, compliance, and enforcement depend on static conceptions of what norms are, and as a result they have blind spots in terms of identifying certain dimensions of change in the social interactions of actors.[46]

Moderate social constructivists emphasise the 'inherently social and intersubjective' nature of social learning.[47] This is in contrast to other kinds of scholarship, particularly ones that are driven by rational choice theories, which focus on changes in the preferences of actors.[48] For instance where an actor like the state 'is not able to generate a satisfactory response' to its constituency as an organisation it may, to adapt, seek to 'get ideas from another thoughts world' which might be the international community.[49] Such a view discounts the potential constitutive role of ideas and structures of knowledge in relation to how actors engage internationally.[50] In other words the realist and liberal approaches to international relations which rely on rational choice theory see learning as a response of actors to a stimulus or change in their environment.[51] This limits how much rational choice theories can explain about social learning within communities and collective enterprises.

In contrast, the social constructivist literature in international relations sees social learning as dynamic, emergent and heavily dependent on socialisation, social influence and, as described below for the purposes of this work, common and collective knowledge otherwise known as intersubjectively developed structures of knowledge and ideas. Within

---

[44] Adler, above n 14, 20. For the beginnings of the approaches to constructivism and learning see in particular the work of Jean Piaget, *The Moral Judgment of the Child* (1932).

[45] See chapter 3 of this book.

[46] For an incisive critique of the concept of obligation and its use to describe the bindingness of norms in international law see Christian Reus-Smit, 'Politics and International Legal Obligation' (2003) 9(4) *European Journal of International Relations* 591.

[47] Thomas Risse, 'Global Governance and Communicative Action' (2004) 39(2) *Government and Opposition* 288.

[48] For instance, Jack Levy, 'Learning and Foreign Policy: Sweeping a Conceptual Minefield' (1994) 48(2) *International Organization* 279.

[49] Farkas, above n 14, 48.

[50] James Rosenau, *The Study of World Politics: Globalization and Governance (Vol 2)* (2006) 115–119.

[51] Adler, above n 14, 19. In the context of theories on learning see Wenger, above n 16, 279.

the social constructivist literature, however, one can differentiate between approaches to social learning depending on what one is looking at or trying to examine.[52] One approach to social learning focuses on 'individual social-psychological changes', whereas another assesses the 'collective learning process that constitutes the practices of social and political communities'.[53] The former is predominantly concerned with the way in which the interaction of actors with each other or socialisation generally changes their beliefs and identities.[54] In these studies, norms are examined for their ability to change the beliefs and identities of actors in terms of whether and how they are internalised.[55] The aim of this line of inquiry is to better understand how norms, ideas and other intersubjective structures are adopted and internalised by individual actors like states.[56] Social learning takes place because an actor is effectively persuaded to adopt particular views. Alternatively, through their interaction with other actors they are socially influenced to respond to issues in particular ways.[57]

Wendt, whose ideas on common and collective knowledge were described earlier,[58] discusses social learning from the perspective of the individual actor using theories on symbolic interactionism.[59] His conception of social learning requires other actors to be present and for interaction to take place between them and others. However, through this interaction with others an actor better understands who they are and their interest in the light of those who are around them. This description of social learning however does not explain the instantiation or cognitive evolution of particular cultures within the community or the collective life of actors

---

[52] The term socialisation is often used in constructivist approaches to international relations to highlight how the interaction of actors with each other shapes how and what they learn. In this sense this book draws from a variety of different approaches to social learning within the constructivist tradition in international relations. For instance, see Alastair Johnston, 'Treating International Institutions as Social Environments' (2001) 45(4) *International Studies Quarterly* 487; Checkel, above n 5; Richard Price and Nina Tannenwald, 'Norms and Deterrence: The Nuclear and Chemical Weapons Taboos' in Peter Katzenstein (ed), *The Culture of National Security: Norms and Identity in World Politics* (1996) 114.

[53] Adler, above n 14, 20–21. For a study of change in the beliefs of individual actors through social learning see Farkas, above n 14; Wendt, above n 3, particularly ch 7.

[54] Checkel, above n 42; Wendt, above n 3, particularly ch 7.

[55] Alderson, above n 42; Flockhart, above n 42.

[56] Adler, above n 14, 20; Finnemore, above n 43; Tannenwald, above n 43.

[57] For instance, Flockhart, above n 42; Johnston, above n 52.

[58] See section 2.2.1.

[59] Wendt, above n 3, 326–336. On symbolic interactionist theories see George Herbert Mead, *Mind, Self, and Society* (1934).

more generally. It is more concerned with the way in which the individual constructs mental structures or learns to value a particular norm.[60]

Such approaches to social learning focus on the 'change of beliefs at the individual cognitive level, either in relation to values, norms, procedures or new routines'.[61] These scholars have a tendency to focus more on the socialisation of actors into groups and particular ways of doing what is already established as the intersubjectivity within particular social contexts.[62] A focus on socialisation, in other words, presumes that particular background knowledge structures already exist and are being practised by at least one other actor within a setting or a context.[63] It also means that individual behaviour is likely to change and will comply with 'the standards set by the group the agent is being socialized into'.[64]

It is arguable that conceptions of learning which focus only on what the individual actor learns, even if this is set in the context of socialising with others, pay little attention to the collective nature of learning in international relations.[65] The alternative approach adopted in this work is more concerned with collective or communitarian approaches to learning.[66] In

---

60 This is not to suggest that Wendt is not interested in change at the level where actors collectively identify with each other. He refers to this however as structural change within the international system which has a supervenient effect on actors. This is because change at this level requires a change in the collective identity of actors. Structural change requires that actors become interdependent, share a common fate, and be homogenous or alike in what they identify with. Given these conditions, Wendt indicates that actors must act with self-restraint towards others to ensure that structural change is possible: see Wendt, above n 3, 336–366. Wendt's discussion of structural change presumes a much deeper degree of social learning than is required or needed for this study. Social learning does not depend only on actors collectively identifying with each other in a structural sense. Identifying with each other, in other words, does not have to be deeply ingrained in the collective life of actors but can simply be a change in how commonly to approach a particular activity.

61 Trine Flockhart, '"Masters and Novices": Socialization and Social Learning through the NATO Parliamentary Assembly' (2004) 18 *International Relations* 361, 366, referring to Levy, above n 48, 287.

62 See, for instance, Risse, Ropp and Sikkink (eds), above n 42; Checkel, above n 42; Alderson, above n 42.

63 Flockhart, above n 61, 366.

64 Ibid.

65 For an example of this see Farkas, above n 14.

66 Adler, above n 14. Adler draws from Wenger's ideas on communities of practice to develop his conception of cognitive evolution as part of communitarian approaches to social learning. In relation to communities of practice see: Jean Lave and Etienne Wenger, *Situated learning: Legitimate peripheral participation* (1991); Wenger, above n 16; Etienne Wenger, Richard McDermott, and William

social constructivism it is common to talk about social structures as a way of describing how actors identify with each other in terms of how they view their common or collective social reality.[67] So, for instance, there is nothing natural about the term sustainability but it means something only in the context of the social environment within which it has developed and continues to be used. Its broad meaning potentially structures the social relations of actors as they seek mutually to engage with each other. Some constructivist approaches to structure use the term 'holism' for instance to describe how 'the parts of a whole behave as the whole requires'.[68] It is in this sense that Adler, for example, talks about communities 'not as a matter of feelings, emotions, and affection, but as a cognitive process through which common identities are created'.[69]

A communitarian approach to social learning does not presume an affinity or solidarity amongst actors that simply does not exist or would be viewed as normative in its orientation.[70] In other words, this conception of social learning presumes that the common or collective knowledge, or the

M Snyder, *A Guide to Managing Knowledge: Cultivating Communities of Practice* (2002); Eric Lesser, Michael Fontaine and Jason Slusher (eds), *Knowledge and Communities* (2000). Although this book adopts Adler's conception of social learning it adds to it using Wendt's conceptualisation of common and collective knowledge. This is because Adler's conception of social learning in international relations is still thin in terms of the concepts needed to explain different kinds of background knowledge or cognitions.

[67] For instance, Nicholas Onuf, *World of Our Making: Rules and Rule in Social Theory and International Relations* (1989) ch 1; Wendt, above n 3, ch 4; Ruggie, above n 6, ch 5.

[68] Martin Hollis and Steve Smith, *Explaining and Understanding International Relations* (1990) 5.

[69] Adler, above n 14, 195.

[70] For instance, solidarist conceptions of international society within the English school of international relations presume that the community is more than the sum of what individual states need to co-exist with each other: see Andrew Hurrell, 'Society and Anarchy in the 1990s' in Barbara Allen Roberson (ed), *International Society and the Development of International Relations Theory* (1998) 17, 27–29. Thomas Franck in his discussion of fairness in international law also presumes that states are part of a community, but one which he sees as having the potential to come to terms with values of its own in a similar way to that of the solidarist view of the English School; see Thomas Franck, *Fairness in International Law and Institutions* (1995) particularly ch 1; Dino Kritsiotis, 'Imagining the International Community' (2002) 13(4) *European Journal of International Law* 961. For an incisive and powerfully argued piece discussing the significance of the idea of community in the development of international law see Bruno Simma and Andreas Paulus, 'The "International Community": Facing the Challenge of Globalization' (1998) 9 *European Journal of International Law* 266.

background knowledge or cognitions of the community are also continuously changing and becoming something else, but that they are an intersubjective understanding of reality which actors share.[71] In other words it is the intersubjectivity amongst actors that emerges and shifts as they engage with each other. The conception of social learning discussed here does not suggest that the community or what it understands as its intersubjectively shared ideas is ontologically different from or beyond what the individual actor wants or needs in any particular way.[72] This is why the concepts of common knowledge described above give meaning to how the interaction of actors with each other have a structure that is different from the collective social structure that was also described above. In other words, common knowledge shifts and develops in what it may mean over time, but it can be captured as an aspect of the way that actors see their immediate reality in a particular moment of time. This intersubjectivity is in itself an aspect of the kind of social learning which is communitarian rather than purely based on what the rational individual actor does.

In this sense, social learning is about the emergence of common knowledge amongst a group of actors who mutually engage with each other.[73] Common knowledge as described above is intersubjectively shared amongst actors because their interaction (direct and indirect) with each other helps them to negotiate meaning out of a particular context or a situation. In this sense, social learning or the emergence of common knowledge is not as deeply transformative as social structures through total collective identification of actors with each other. This is because the interlocking beliefs and understandings of actors can shift with new circumstances and events.

On the other hand, structural change or the emergence of collective knowledge is more deeply transformative but harder to achieve.[74] Wendt points out that structural change 'occurs when actors redefine who they are and what they want'.[75] An example of collective knowledge for instance is in the way actors within particular international communities

---

71 See also Adler, above n 14. Adler does not use the term 'common or collective knowledge'.

72 Ibid, 13.

73 For an example of what this means in the context of this work see for instance chapter 6 of this book which discusses the diffuse and indirect expression of productive power by the International Court of Justice.

74 For instance, Wendt argues that collective identification requires interdependence, common fate and homogeneity to be present amongst actors actively to facilitate structural change. In addition, actors themselves must exercise self-restraint to ensure that structural change happens: Wendt, above n 3, 343–366.

75 Ibid, 336.

currently collectively identify with the philosophical ideals of protecting the environment, not just because of its immediate benefits to all who currently inhabit the earth but also for those who are yet to be born.[76] In other words, why we protect the environment in the first place is in itself a socially constructed idea which is taken for granted in most international engagements of actors with each other.[77] Actors' interests are determined by the way they collectively identify with others in terms of needing to show self-restraint in relation to protecting the environment. This does not say much about the depth of their commitments in particular instances, but actors rarely pick the indiscriminate exploitation of the environment as an alternative when in dialogue with others internationally. In this sense social learning also occurs when actors instantiate a culture that is collectively shared or is the social structure that determines who they are in their relationships with each other and those not within their communities. For instance, sovereignty as an idea that states all collectively identify with is regularly instantiated when the United Nations meets and gives all non-state actors a status as observers. Importantly, it is the collective or the community of states that is continuing to learn about the instantiations of an existing collective identity which differentiate this learning experience from the individually based one discussed earlier.

The possibility that social learning happens at the level at which actors organise themselves into institutions or communities is important for this work. This is because common knowledge or the instantiation of collective culture as components or expressions of what has been learnt by communities can be identified through research without having to examine how the individual preferences of actors or their identities have also changed. This is not to suggest that agency or the individual actor is not relevant, but that social learning expressed as reifications of intersubjectivity can be examined in scholarly works without having to ascertain the extent to which the individual actors' interests and identities have also changed. A focus on collective social learning however says nothing about how it happens and the role and function of environmental principles in that process. Some approaches to social learning would assume that the

---

[76] See DesJardins, above n 31; Dryzek, above n 4, 47–49; Dryzek, above n 32.

[77] Note that one of the main arguments of chapter 6 is that transnational corporations are socially learning by instantiating the idea that the future generation must have the environment protected for it in ways that are framed by the environmental principles of the Global Compact. In other words, the idea that the environment must be protected cannot be taken for granted, but yet it is a structural dimension of the relationship of states with each other even if they differ in terms of what this means in particular contexts.

consent of an actor to a norm or the entry into force of an international agreement would naturally help with how an actor learns.[78] The discussion below assesses how collective learning happens because of social processes that steer the cognition of communities. The particular kind or the nature of these processes important for steering learning in particular directions is also discussed in chapters 6, 7 and 8, where they receive closer attention in the context of the case studies in this work.

2.3.1 Social Practice, Social Processes and Collective Learning

The fact that moderate social constructivists do not discuss or critically engage with the 'microprocesses' or mechanisms that help us to study how institutional facts or intersubjective structures create meaning for actors has been criticised by some international relations scholars.[79] This means that the focus is often on the supposition that social structure constitutes actors rather than examining their social practices or processes that allow us to understand the interplay between reifications of various kinds and the participation of actors in social structures.[80] The focus is often on structure and not the micro and agency level processes that make the interplay between structure and agent meaningful.[81] Ringmar has made this point in his criticism of Wendt's version of constructivism discussed above:

> His structural bias constantly sets limits to his investigation: just as the structure of the international system cannot make a state act, it cannot make someone have a particular identity. Just as the structure cannot explain historical change, it cannot by itself explain changes in identities. What Wendt needs, but cannot provide with the help of the theoretical perspective he has made his, is an account of how states interpret the structures of international politics and how they use them in interaction with others.[82]

---

[78] See for instance Adler's work which uses the terms coined by Karl Popper to call this the 'bucket' view of learning: Adler, above n 14, 19. This means that a community, a collective group of actors or even individuals are presumed to be like buckets which simply learn whatever is 'taught' to them.

[79] See, for instance, Johnston, above n 52; Jeffrey Checkel, 'The Constructivist Turn in International Relations Theory' (1998) 50 *World Politics* 324; and, more recently, Flockhart, above n 42. In terms of how this has been done in international relations see for instance the 2005 special edition of *International Organization*, volume 59(4), put together by Jeffrey Checkel and later edited as a book: Jeffrey Checkel (ed), *International Institutions and Socialization in Europe* (2007).

[80] The terminology linking participation and reification is borrowed from Etienne Wenger; see Wenger, above n 16, ch 1.

[81] See Checkel, above n 5.

[82] Erik Ringmar, 'Alexander Wendt – a Social Scientist Struggling with

It seems that this is another way of saying that structure alone does not explain how actors negotiate meaning as they engage with others internationally. This supposed ambivalence towards the role of the agent in creating structures in the first place is partly due to the ease with which, in applied studies, one can confuse constructivism with rational choice theories and their approaches to how actors form interests in something.[83] However, it is also easy to forget that underpinning rational choice theory is the presumption that the agent is rational and can act in this way across a variety of different situations or cases.

Increasingly, the scholarship on social constructivism has focused on the social practices of actors as a way of identifying the real meaning that structures have for them in the context of their interactions with others.[84] This focus on practice is meant to highlight how the activities of actors internationally are 'socially situated' rather than as solely strategic or instrumental. A focus on social practice has two interrelated objectives. It allows us to identify how the participation of actors within the community is constituted by the intersubjectivity that exists within it. However, it also enables an examination of how the social practices of actors that are constituted by ideas, knowledge, norms, or other cognitive frames or reification find expression in reality *which reinforces them or variations of*

History' in Iver Neumann and Ole Waever (eds), *The Future of International Relations: Masters in the Making* (1997) 269. Similar comments have also been made by other scholars. Flynn and Farrell write that:

> instead of fully exploiting the power of the insights they borrow from social theory about the recursive nature of the relationship between agent and structure, constructivists have ended up seeking to demonstrate only that norms as elements of structure (alongside material conditions) can determine the interests and identity of agents, rather than seeking to locate the power of norms in the process whereby they are created in the first place.

Gregory Flynn and Henry Farrell, 'Piecing Together the Democratic Peace: The CSCE, Norms, and the "Construction" of Security in Post-Cold War Europe' (1999) 53 *International Organization* 505, 510–511.

This is another way of saying that social constructivist approaches lack a theory of the agent or the actor who is regulated or constituted by structure. On this see Checkel, above n 79; and Mark Harvey, 'Developing the Agent: Towards a Constructivist Theory of the Identity' (Paper presented at the British International Studies Association Conference, Edinburgh, 2000).

83 Those who write from a social constructivist perspective in international relations indicate that it has no particular theory about the agent. See Flockhart, above n 61; Flockhart, above n 42.

84 See, eg, work of Antje Wiener, 'Contested Compliance: Interventions on the Normative Structure of World Politics' (2004) 10(2) *European Journal of International Relations* 189; Adler, above n 14.

them. In this way, social practice is both a confirmation of existing cognitions and an expression of it in practice, thereby highlighting the learning that is going on. Instead of having a theory of how the agent acts, the focus on social practices forces one to study the way in which intersubjectivity is made meaningful in one context in which it may look different from how it has been expressed in a different context. Therefore, reifications like an environmental principle depend on the social practices of actors for expressing it in a particular context. It also means that the common knowledge or the interlocking beliefs of actors in two different areas of practice will not necessarily be the same.

A focus on the social practices of actors is another way of referring to the potential material expression of, or doing, what is collectively acceptable and understood (for example, rules) and which gives actors their bearing in their social relations with others.[85] In this sense, their practices are the property of the communities and not the individuals because of the 'sustained pursuit' of activities or enterprises that actors have shared with each other.[86] Their ideas of how to pursue particular activities within the international context are driven by the practices of their communities which are also theirs. It is the expression of that intersubjectivity in the particular context, however, that makes things meaningful and highlights the social learning going on.

The concept of practice is based on the idea that actors are always negotiating the meaning of their participation within a particular community or group of people who mutually engage with each other in some way. The intersubjectivity that emerges from the participation of actors is what they use to engage in meaningful practices within their community or institutional context. It is the active participation of actors within a context which helps the group not only to develop its cognition and to learn socially but to reinforce the reified expressions of intersubjectivity. A rule, whether formal or informal, is made meaningful and relevant through the social practices of the actors that seek to use it as they mutually engage with others.

A focus on practice is important for understanding social learning and also how the agent or actor reflexively responds to structures of thought within a community or organised group of actors. This is consistent with a view that focuses on 'learning-by-doing'.[87] This takes the participation

---

[85] Adler, above n 14, 265. See also Wenger, above n 16, 45.

[86] Ibid, 45.

[87] See Adler, above n 14, 20, who is referring to Daniel Kahneman, Paul Slovic and Amos Tversky, *Judgment under uncertainty: Heuristics and biases* (1982). The view is an alternative to what Adler has called the 'bucket view of learning',

of actors in the learning process more seriously and assumes that it is integral to change or the way in which they come to construct meaning for themselves in the context of the prevalent forms of intersubjectivity.[88]

It is also consistent with the views expressed by Wiener, who examines the differences in approach between what norms as structures mean to actors.[89] Her argument is that the constructivists usually see norms as a kind of structure either regulating or constituting actors.[90] In this sense, actors simply react to norms or structure, whereas a 'reflexive understanding' of the dynamic within groups interacting with each other 'implies that the meaning of norms as the dependent variable is embedded in social practice'.[91] This view appears to reinforce the 'learning-by-doing' approaches to social learning because it suggests that for norms to have real impact within communities or groups one needs to examine them in the context of how actors mutually engage with each other in their practices.[92] This particular approach also departs from seeing social learning as the socialisation of individual actors into having particular interests or beliefs. That is, the background knowledge structures for groups of actors develop organically because of the reflexive relationship they have with structures of thought such as norms.

Charles Taylor, writing about social practice, has put the idea very simply by noting that:

> By 'practice', I mean something extremely vague and general: more or less any stable configuration of shared activity, whose shape is defined by a certain pattern of dos and don'ts, can be a practice for my purpose. The way we

---

whereby one presumes that by giving actors information they simply learn it and internalise it.

88 Ibid, 20.

89 See, for instance Wiener, above n 84; Karin Fierke and Antje Wiener, 'Constructing institutional interests: EU and NATO enlargement' (1999) 6(5) *Journal of European Public Policy* 721. On the idea of social practice see more generally: Anthony Giddens, *Central Problems in Social Theory: Action, Structure, and Contradiction in Social Analysis* (1979); Pierre Bourdieu, *Outline of a Theory of Practice* (1977); Charles Taylor, 'To Follow a Rule . . .' in Craig Calhoun, Edward LiPuma and Moishe Postone (eds), *Bourdieu: Critical Perspectives* (1993) 45.

90 Wiener, above n 84, 191.

91 Ibid.

92 See also the following comment from Taylor that Antje Wiener, above n 84, uses in her work to describe the approach to practice: 'the practice not only fulfils the rule, but also gives it concrete shape in particular situations. Practice is a continual "interpretation" and reinterpretation of what the rule really means': Taylor, above n 89, 57.

> discipline our children, greet each other in the street, determine group decisions through voting in elections, and exchange things through markets are all practices. And there are practices at all levels of human social life: family, village, national politics, rituals or religious communities and so on.[93]

In a different work, Taylor also highlights that a focus on practice 'not only fulfils the rule, but also gives it concrete shape in particular situations. Practice is . . . a continual "interpretation" and reinterpretation of what the rule really means.'[94] In the context of international politics, Wiener uses the term social practice to mean discursive interventions, which include things like 'official documents, policy documents, political debates and media contributions'.[95] What Wiener appears to ignore in her discussion of social practice is the way that discursive interventions, as per the examples she gives, are deeply constituted and dependent on persuasion and social influence as factors that socialise actors or make up their mutual engagements with each other. In other words, discursive practices of actors are still dependent on factors like persuasion, social influence, and power.[96] However, it is through their social practices that one can better identify the meaning that a particular structure, such as norms, has for a group or community of actors. The expressions of common knowledge through social practice highlight the collective social learning that is going on within the group or the community of actors.

2.3.2 Social Process and Collective Social Learning

This focus on social processes is an alternative way of examining social learning as opposed to what scholars from a constructivist perspective have been focusing on in their studies of socialisation. A number of scholars writing from a social constructivist perspective have been focusing on the mechanisms that socialise actors into structures or ways of changing their behaviour or beliefs in response to ideas, norms, values, et cetera.[97] Flockhart has summarised two strategies well:

---

[93] Charles Taylor, *Sources of the Self: The Making of the Modern Identity* (1989) 204.

[94] Taylor, above n 89, 57.

[95] Wiener, above n 84, 201.

[96] See for instance the collection of papers in David Barton and Karin Tusting (eds), *Beyond Communities of Practice: Language, Power and Social Context* (2005).

[97] See, eg, Checkel, above n 42; Flockhart, above n 42; Rodger Payne, 'Persuasion, Frames and Norm Construction' (2001) 7 *European Journal of International Relations* 37; Finnemore and Sikkink, above n 42; Risse, above n 47;

> 'Social influence', which elicits pro-norm behaviour through the distribution of social rewards and punishments, and 'persuasion', which encourages norm consistent behaviour through a social process of interaction that involves changing attitudes without use of either material or mental coercion.[98]

The idea that persuasion is an effective tool for understanding why actors might change their beliefs and identities is commonly discussed in the social constructivist literature evaluating why norms, as social structure or reifications, can be effective.[99] Finnemore and Sikkink use persuasion and their version of social influence to argue that they are steering mechanisms which can be used within particular contexts to change the behaviour of other actors towards accepting the logic of what is normatively appropriate.[100] The difficulty with the concept of socialisation, as described above, is that it takes intersubjective structures as formed and established. As a result, persuasion, for instance, is a mechanism that can socialise actors into changing their beliefs or identities. It does not, however, give us any particular insights into whether the participation of actors within a group makes a difference in terms of creating new common knowledge amongst them that is reflective of a particular frame or structure like a norm such as an environmental principle. This does not mean that the concepts of persuasion or social influence are not useful in the context of this work, but the way in which they are used as a mechanism by some constructivists does not bring out their potential for collective social learning.

Combing the various discussions above, this section has tried to deal with an important question relating to how we can meaningfully analyse what is happening in communities, groups, and institutions in order to assess whether there has been collective social learning or not. In other words, just because we see a rule which has received the consent of states it does not mean that it is something that they intersubjectively share in terms of a common knowledge of what is the appropriate thing to do. This is why, as described above, international relations scholars have also found

---

Alastair Johnston, 'Conclusions and Extensions: Toward Mid-Range Theorizing and Beyond Europe' (2005) 59 *International Organization* 1013.

[98] Flockhart, above n 42, 97. There is a wealth of scholarship discussing the role of persuasion in international relations. For a good examination of this in the context of international institutions see Johnston, above n 52; Finnemore and Sikkink, above n 42, 902–903.

[99] Checkel, above n 42; Flockhart, above n 42; Payne, above n 97.

[100] They refer to social influence as conformity and esteem which they say involves 'evaluative relationships between states and their state "peers"': Finnemore and Sikkink, above n 42.

it useful to describe how individual actors can be socialised through, for instance, persuasion or social influence to internalise particular norms.

A focus on the social practices of actors highlights situations where real social learning has happened because of the way actors are expressing what they intersubjectively believe to be their reality as a group. So, for instance, actors engage in a practice when they have to report on their activities to an organised group of actors. The extent to which the practice of reporting, which is carried out individually, is structured in a particular way indicates that the group is learning something. However, the depth of this learning experience can be quantified only if one is able to identify whether it is reflective of some structure of thought of some particular kind. This is where the language of common knowledge and the instantiation of collective culture help. However, the social practices of actors do not explain why they have learnt as a group to reflect a common knowledge of a particular kind. This is where social influence or persuasion becomes a part of the collective learning experiences of a group of actors. To the extent to which they seek to apply to themselves what someone else is doing well, in the context of what is appropriate for the group to do, they are being socially influenced by others to think in terms of the common knowledge in that group.

The implications of the above discussion are that collective social learning needs to be looked at in terms of how it is reflected in the social practices of actors. However these social practices need to be contextualised in terms of what is driving or steering the learning process for individual actors themselves who are part of the community or group.[101] It is here that the description and work of social constructivist scholars who have examined socialisation in international relations are helpful. This is because they have already carried out work in terms of how individual actors can internalise the intersubjectivity that is already developed. Actors might mutually engage each other to either bargain or argue about something that is important to all of them.[102] As a result they may co-exist or co-operate depending on whether they have bargained or been persuaded by the logic of the arguments put forward by all those who were engaging with each other. However, unless they are persuaded by the arguments of each other it is difficult to say that they have collectively learnt because the concept of common knowledge requires interlocking beliefs. As a result

[101] For instance, Kerstin Jacobsson, 'Soft Regulation and the Subtle Transformation of States: The Case of the EU Employment Policy' (2004) 14(4) *Journal of European Social Policy* 355; Risse, above n 47.

[102] See chapter 8 of this book.

the practices of actors who have bargained will look as if their knowledge has emerged out of mutual engagement or the social interactions of the group but the logic of what is appropriate for them will disappear as soon as the incentives have gone.

2.4 CONCLUSION

This chapter has described in theoretical terms the social constructivist approach that this work will take to understand the role and function of environmental principles in international politics. Using this theoretical framework this chapter has conceptually differentiated between 'social learning' and 'socialisation' even though constructivists are concerned with both. It described the term 'social learning' as something that actors do when they create common knowledge or instantiate collectively held cultures within their group. Individual actors on the other hand are socialised into learning new identities and beliefs consistent with what is intersubjectively shared amongst an already existing group of actors. This difference is important because the question driving this work seeks to examine the role and function of environmental principles in terms of framing social learning and change in international law and politics. As a result, it is not concerned with how an individual is socialised to learn how to conform to environmental principles.

Importantly, this chapter has also highlighted how we can observe whether collective social learning happens or not. This is the more complex part of this discussion, because whether there is social learning is intimately intertwined with how we observe it. The discussion of social practice is meant to highlight how structures of thought or intersubjectivity are always being confirmed and created at the same time as actors practise what they believe to be appropriate within their community or group. So, for instance, the term sustainable development has a particular intersubjectively held meaning for the group of actors whose practices we are looking at. Through their social practices we can identify what they commonly hold as their beliefs about what environmental principles mean to them in a particular context.

More importantly, the participation of actors or their interactions with each other do not of and in themselves bring about change. Actors create meaning from reifications, whether they be norms or something else, by participating in the processes that construct meaning from them. In groups, intersubjectivity develops because certain social processes drive or steer the construction of meaning during interactions that actors have with each other. It is these social processes that are discussed in chapters 6,

7 and 8 as the mechanisms which steer or drive social learning and which explain the functional potential of abstract and open-textured norms like environmental principles. This chapter discussed the relationship between the idea of social processes and practice to explain how meaning and function are conditions of particular contexts. In the later chapters 6, 7 and 8 the different social processes that steer learning are described in more detail.

PART II

Norms, environmental principles and social learning

3. Environmental principles as abstract or open-textured norms

3.1 INTRODUCTION

In international law and politics, principles are often defined by their source, in terms of how they function, or whether they are linguistically structured in a different way from other norms like rules or standards. For instance, the *Statute of the International Court of Justice* was drafted by creating a conceptual category known as 'general principles of law' which, when recognised as such by civilised nations around the world, would also be a source of international law.[1] The concept of a principle is also commonly used to refer to certain particular types of norms that are recognised in customary international law or are drafted into international agreements and resolutions of international organisations and institutions.[2] Where a principle comes from, however, does not say much about what it is and if it is capable of doing anything in terms of changing how actors at the international level commonly and collectively develop interests in relation to something.

In the context of international law and politics the term 'principle' is used in many different ways, but many approaches to describing principles rely on how they are functionally or structurally different from rules.[3] This

---

1 See *Statute of the International Court of Justice*, Art 38(1)(c). On general principles of law as a source of international law see, for instance, Olufemi Elias and Chin Lim, '"General Principles of Law", "Soft Law" and the Identification of International Law" (1997) 28 *Netherlands Year Book of International Law* 3; Christopher Ford, 'Judicial Discretion in International Jurisprudence: Article 38(1)(c) and "General Principles of Law"' (1994) 5 *Duke Journal of Comparative and International Law* 35; Frances Jalet, 'The Quest for the General Principles of Law Recognized by Civilized Nations – A Study' (1963) 10 *UCLA Law Review* 1041.

2 For general studies in international law of the concept of a principle see, Vladimir Degan, *Sources of International Law* (1997) especially ch II; Geza Herczegh, *General Principle of Law and the International Legal Order* (1969); Georg Schwarzenberger, *The Fundamental Principles of International Law* (1956).

3 Michael Moore, 'Legal Principles Revisited' (1996–1997) 82 *Iowa Law Review* 867, 870.

is also the case in terms of environmental principles, with many scholars adopting the distinction that Dworkin has developed between principles and rule.[4] He has written:

> The difference between legal principles and legal rules is a logical distinction. Both sets of standards point to particular decisions about legal obligation in particular circumstances, but they differ in the character of the direction they give. Rules are applicable in all-or-nothing fashion. If the facts a rule stipulate are given, then either the rule is valid, in which case the answer it supplies must be accepted, or it is not, in which case it contributes nothing to the decision.[5]

He then argues that principles merely give 'a reason that argues in one direction but does not necessitate a particular decision'.[6] Dworkin's ideas have developed from the assumption that legal principles which are not written are reflective of what courts and adjudicative processes do if one collated the variety of rules that they seek to apply to given situations.[7] The difficulty with juxtaposing the views of Dworkin in relation to environmental principles at the international level is that his conception of a principle requires enough social practice amongst judges and other actors more generally for principles to emerge. The difficulties in identifying customary international law, for instance, make the task of seeing principles in this way rather problematic.[8]

It is also difficult to see how the same environmental principle can

---

[4] Jonathan Verschuuren, *Principles of Environmental Law: The Ideal of Sustainable Development and the Role of Principles of International, European, and National Environmental Law* (2003) 38; Ulrich Beyerlin, 'Different Types of Norms in International Law' in Daniel Bodansky, Jutta Brunnée and Ellen Hey (eds), *The Oxford Handbook of International Environmental Law* (2007) 425; Philippe Sands, *Principles of International Environmental Law* (2nd ed, 2003) 231; Lavanya Rajamani, 'The Principle of Common but Differentiated Responsibility and the Balance of Commitments under the Climate Regime' (2000) 9 *Review of European Community and International Environmental Law* 120, 124. This is understandable, given the praise Dworkin has received, even amongst his critiques, for the rigorous way that he has developed his distinction between legal and moral principles as well as rules and principles. See Larry Alexander and Ken Kress, 'Against Legal Principles' (1996–1997) 82 *Iowa Law Review* 739, 745–752. For a good example of someone writing about the concept of the environmental principle without using Dworkin's terminologies see Nicolas de Sadeleer, *Environmental Principles: From Political Slogans to Legal Rules* (2005).

[5] Ronald Dworkin, *Taking Rights Seriously* (revised ed, 1978) 24–25.

[6] Ibid, 26.

[7] Moore, above n 3. See also Alexander and Kress, above n 4.

[8] See for instance the arguments against the principle of common but differentiated responsibilities in the work of Christopher Stone, 'Common but

provide similar reasons for action across different social contexts, given the number of different ways that some of them are described or codified in various international agreements. By defining a principle in this way, Dworkin potentially removes the role of participation and interaction within social contexts from generating not only the meaning that particular principles have for actors, but also how they function in creating change for them as a group. The interplay between the participation of actors in groups, or the social processes that drive change within them, and an open-textured principle means that their function has to be ascertained through the social practices of actors. Principles may either constrain or stimulate new opportunities to arise for actors to learn how collectively to identify with each other. These possibilities need to be examined by paying attention to the practices of actors within groups. Dworkin's conception of a principle as reason for action could be relevant if the concept of an environmental principle were being examined in the context of judicial decision-making. Given that the focus of this work is more broadly conceptualised in terms of social learning, what is needed is a definition of a principle which is able to accommodate the variable nature of participation in different social contexts.

Not everyone collapses the functional and linguistically structured differences between rules and principles into the one description. Braithwaite and Drahos, drawing on Raz, define principles simply in terms that describe how precisely or abstractly they are drafted in relation to standards and rules.[9] However, just because norms are defined or contrasted against rules in terms of being open-textured says nothing about how their role or function might be interpreted in social contexts. Kratochwil has identified some norms as enabling 'parties whose goals and/or strategies conflict to sustain a "discourse" on their grievances, to negotiate a solution, or to ask a third party for a decision on the basis of commonly accepted rules, norms, and principles'.[10] In contrast, however, Ruggie has referred to 'generalised principles of conduct' as 'principles which specify appropriate conduct for a class of actions, without regard to the

---

Differentiated Responsibilities in International Law' (2004) 98 *American Journal of International Law* 276.

[9] John Braithwaite and Peter Drahos, *Global Business Regulation* (2000) 18–20. At 9, Braithwaite and Drahos define principles as 'abstract prescriptions that guide conduct'. On Joseph Raz and principles see in particular: Joseph Raz, 'Legal Principles and the Limits of Law' (1971–1972) 81 *Yale Law Journal* 823.

[10] Friedrich Kratochwil, *Rules, Norms, and Decisions: On the Conditions of Practical and Legal Reasoning in International Relations and Domestic Affairs* (1991) 70.

particularistic interests of the parties or the strategic exigencies that may exist in any specific occurrence'.[11] The differences between Kratochwil's and Ruggie's definitions of a norm and a principle appear to lie in the way they value their function. Whereas Kratochwil sees principles as problem solving devices, Ruggie defines them as guiding behaviour. Kratochwil's definition builds the abstractness and open-textured quality of a principle into the functional capacity it has for solving problems, whereas Ruggie simply defines a principle as being more open-textured than a rule. That is, in Ruggie's case both rules and principles guide behaviour, whereas in the case of Kratochwil a principle is functionally different from rules in the way it structures how political relations are determined. Both scholars are however interested in the social impact of norms that are indeterminate or open-textured in terms of how they are structured.

Although Kratochwil and Ruggie both write as social constructivists in international relations, the general literature in this field discussing norms in international law and politics does not generally make an issue out of the degrees of abstraction, ambiguity or vagueness in norms.[12] Constructivists seem less interested in the possibility that the way norms as structures of intersubjective meaning or episteme might be reified in different ways.[13] For instance, Flockhart, while summarising the views of a number of social constructivists in international relations, has written that 'norms tend to be stable structures acting as constraints on agents' behaviour and as constitutive of identity and interests'.[14] This definition

---

[11] John Ruggie, 'Multilateralism: The Anatomy of an Institution' in John Ruggie (ed), *Multilateralism Matters: the Theory and Praxis of an Institutional Form* (1993) 3, 11.

[12] For exceptions see, for instance, Antje Wiener, 'Contested Compliance: Interventions on the Normative Structure of World Politics' (2004) 10(2) *European Journal of International Relations* 189; Christian Reus-Smit, 'Human Rights and the Social Construction of Sovereignty' (2001) 27 *Review of International Studies* 519; Rodger Payne, 'Persuasion, Frames and Norm Construction' (2001) 7 *European Journal of International Relations* 37.

[13] Although constructivists in international relations and politics point out that norms have an ontologically independent status. On this see, for instance, Martha Finnemore and Kathryn Sikkink, 'International norm dynamics and political change' (1998) 52(4) *International Organization* 887; Annika Björkdahl, 'Norms in International Relations: Some Conceptual and Methodological Reflections' (2002) 15(1) *Cambridge Review of International Affairs* 9; Jeffrey Checkel, 'The Constructivist Turn in International Relations Theory' (1998) 50 *World Politics* 324.

[14] Trine Flockhart, '"Complex Socialization": A Framework for the Study of State Socialization' (2006) 12(1) *European Journal of International Relations* 89, 91. Jeffrey Checkel, who writes using social constructivism, has for instance

is consistent with the more commonly cited ones adopted by social constructivists which state that they are 'collective expectations about proper behaviour for a given identity'.[15] This general approach to norms appears restrictive, given that it presumes that they are not likely to be ambiguous and that their meaning and real function for actors does not emerge out of the interaction of actors with each other and within particular social contexts. Given that environmental principles are generally expressed, as will be argued below, as open-textured norms or as abbreviations of abstract ideas, this chapter discusses what this means in terms of their potential to engage actors in a way that creates meaning for the participants or enables them to learn as a group. It discusses how the meaning and function of environmental principles must be examined in the context of the discursive and other social practices of actors because they are open-textured and abstract. This also suggests that arguments about whether they mean anything are more productively examined in terms of whether they are able to interplay with or constitute the participation of actors in social contexts that they function in. In this way, this chapter applies existing social constructivist approaches to explain the intersubjectively developed nature of norms, but extends this by arguing how their role in terms of what they mean and how they function emerges out of the social practices of actors.

3.2 OPEN-TEXTURED QUALITIES OF PRINCIPLES

To suggest that environmental principles have a structure might seem odd, given that they are often expressed in abbreviated forms. However the precautionary principle, which is almost always referred to in an abbreviated form, is also much more detailed when it is actually codified as a norm.[16]

---

commented that 'for a norm to exist, it thus must embody clear prescriptions, which provide guidance to agents as they develop preferences and interests on an issue': Jeffrey Checkel, 'The Europeanization of Citizenship?' in Maria Green Cowles, James Caporaso and Thomas Risse (eds), *Transforming Europe: Europeanization and Domestic Change* (2001) 180, 183.

[15] Ronald Jepperson, Alexander Wendt and Peter Katzenstein, 'Norms, Identity, and Culture in National Security' in Peter Katzenstein (ed), *The Culture of National Security. Norms and Identity in World Politics* (1996) 33, 54. See also Finnemore and Sikkink, above n 13.

[16] In relation to the precautionary principles, VanderZwaag has shown how 14 different formulations of it exist in multilateral agreements and declarations: see David VanderZwaag, 'The Precautionary Principle in Environmental Law and Policy: Elusive Rhetoric and First Embraces' (1999) 8 *Journal of Environmental Law and Practice* 355. The version that appears in the *United Nations Declaration*

A working definition adopted to explore the idea of an open-textured environmental principle creates the possibility of a correlation 'between generic cases (sets of properties) and normative solutions' as part of their codified sentence structure.[17] In other words, it is the correlation between the types of situations that make the rule necessary and the prescription which is imposed in various degrees upon that condition.[18] An alternative way to describe this is to distinguish between 'the *scope* of a rule (what fact-situations does it govern?) and its *character* (what kind of prescription?)'.[19] This is referred to as the *protasis* and *apodosis* of a norm.[20]

In this description there is the possibility that both the behaviour that gives rise to requiring a normative solution and the remedy provided for it are openly or abstractly defined and stated. It could also be that the cases that require a normative solution are broadly stated but that the kinds of prescription relating to them are closed in their scope. This distinction between the properties of the case as open and closed categories is important because it can functionally differentiate a principle and a rule. Principles state the *protasis* and *apodosis* of a norm in an open manner. The range of behaviour that requires a normative solution is either broad or undefined or some shade within the range of being open. The normative solution to the open category of behaviour is also potentially defined in an open way or sometimes developed through vague or under-defined concepts and ideas.[21] This means that groups have to participate in negotiating the meaning of the open-ended propositions in principles.

---

on Environment and Development, UN Doc A/CONF.151/5/Rev.1 (1992) as Principle 15 requires that '[w]here there are threats of serious or irreversible damage, lack of full scientific certainty shall not be used as a reason for postponing cost-effective measures to prevent environmental degradation'.

[17] Manuel Atienza and Juan Ruiz Manero, *A Theory of Legal Sentences* (1998) 27. Atienza and Manero draw from Alchourrón and Bulygin's work on norms to create their definitions: see Carlos E Alchourrón and Eugenio Bulygin, *Normative systems* (1971); Carlos E Alchourrón and Eugenio Bulygin, 'The Expressive Concept of Norms' in Risto Hilpinen (ed), *New Studies in Deontic Logic* (1981) 95.

[18] See also William Twining and David Miers, *How To Do Things With Rules* (4th ed, 1999) 132; Gidon Gottlieb, *The Logic of Choice: An Investigation of the Concepts of Rule and Rationality* (1968).

[19] Twining and Miers, above n 18, 133.

[20] In relation to this see ibid, 132–133.

[21] Although see Neil MacCormick, *Legal Reasoning and Legal Theory* (1978) 152; Joseph Raz, *Practical Reason and Norms* (1999) 49, cited in Beyerlin, above n 4, 434. Beyerlin, at 434, refers to MacCormick and Raz as supporting:

> 'the view that rules and principles show only a difference of degree since both are norms that . . . have a relationship of family resemblance with one another', and 'they have a similar or analogical role in legal discretion'.

For example, Principle 25 of the 1992 *United Nations Declaration on Environment and Development* states that '[w]arfare is inherently destructive of sustainable development. States shall therefore respect international law providing protection for the environment in times of armed conflict and cooperate in its further development, as necessary'.[22] In this example, engagement in armed conflict is the type of behaviour that draws a response from Principle 25. It is the *protasis* of the norm. Principle 25 of the *Rio Declaration* also states that the normative condition for protecting the environment during armed conflict is to respect 'international law providing protection for the environment'. This is an open statement because of the range of possible environmental norms that can and may need to be accounted for in a situation that is potentially harmful to all aspects of the natural environment. It may initially seem as if Principle 25 is defined in a potentially precise way because it is about armed conflict and protection of the environment during that conflict. However, it clearly requires a social context for the components of the norm to have more precise meaning for actors. How it contributes to change will vary because of the different ways that the more open portions of the norm can constitute the participation of actors with each other. Importantly, it creates opportunities for actors to refer to a larger body of information or knowledge in deciding what to do.

Miers and Twining suggest that a principle can be categorical in its assertion that a particular normative solution be followed.[23] They use the example of a norm which states that a 'person should not be tortured under any circumstance' and assert that it can function as a principle and a rule because it is open and categorical in its emphasis on the particular prohibition. This is the same as suggesting that an actor 'must' adopt policies that are environmentally sustainable because they are categorical in their assertions. Beyerlin coined the terms 'action-oriented' and 'result-oriented' rules because he saw certain environmental principles as being more categorical in terms of 'making the addressees take action, refrain

---

22 *United Nations Declaration on Environment and Development*, UN Doc A/CONF.151/5/Rev.1 (1992) ('*Rio Declaration*'). For commentary on the *Rio Declaration* see, for instance, David Wirth, 'The Rio Declaration on Environment and Development: Two Steps Forward and One Back, or Vice Versa' (1995) 29 *Georgia Law Review* 599; Ileana Porras, 'The Rio Declaration: A New Basis for International Cooperation' in Philippe Sands (ed), *Greening International Law* (1993) 20; Marc Pallemaerts, 'International Environmental Law From Stockholm to Rio: Back to the Future?' in Philippe Sands (ed), *Greening International Law* (1993) 1.

23 Twining and Miers, above n 18, 126.

from action, or to achieve a fixed result'.[24] According to Beyerlin, Principle 15 of the *Rio Declaration*, which spells out the precautionary principle, is an action-oriented rule.[25] He argues that if the conditions requiring precautionary action are present an actor has little choice but to respond to the potential environmental harm.[26] In these two different formulations of the same idea, using the terms described above, the *protasis* of a norm is open to a range of possibility whereas the *apodosis* is categorical.

Although these views are actually in support of the idea that environmental principles are potentially capable of legal consequences, they also assume that norms simply function only in terms of compelling actors to follow obligations that they have agreed upon. There is also the presumption that the threshold for the application of the supposed rule-like feature of the norms is meaningful, away from the social practices of actors in their particular contexts. Most importantly, in the case of the precautionary principle for instance, to ascribe certainty to the threshold that then requires a response from an actor misunderstands the vocation of the norm.[27] Nicholas de Sadeleer developed the term 'indeterminate rules' to refer to environmental principles like the precautionary principle which give direction to actors but which do not 'have the effect of setting limits to their meaning and preventing them from evolving to meet new contingencies'.[28] This suggests that the determinacy of the principles must be assessed in terms of how, as norms, they constitute social practices within

---

24 Beyerlin, above n 4, 437.

25 See above n 16 for a definition of the precautionary principles as it appears as Principle 15 of the *Rio Declaration*.

26 According to Principle 15 of the *Rio Declaration*, for instance, these conditions include a threat of serious or irreversible damage. Beyerlin acknowledges that a state has discretion to choose how it will respond: Beyerlin, above n 4, 440. The view that Principle 15 is categorical can easily be called into question because nothing in it suggests that the risk of harm must be avoided at any cost: Ellen Hey, 'The Precautionary Concept in Environmental Policy and Law: Institutionalizing Caution' (1991–1992) 4 *Georgetown International Environmental Law Review* 303, 309–310.

27 See, eg, Jaye Ellis, *Soft Law As Topos: The Role of Principles of Soft Law in the Development of International Environmental law* (DCL Thesis, McGill University, 2001) 157.

28 de Sadeleer, above n 4, 309. Sadeleer referred to indeterminate rules as directing principles. See also the following comment by Ellis discussing the precautionary principle:

> Efforts to encapsulate the principle within the four corners of a legal text might have the effect of robbing the principle of its capacity to point to innovative approaches to decision-making and the adoption of environmental measures.

Ellis, above n 27, 160.

the contexts where actors engage with each other rather than in isolation and in abstract terms.

3.3 THE ABSTRACT NATURE OF PRINCIPLES

Not all environmental principles are expressed or employed as norms with a *protasis* and an *apodosis* that are open to a range of interpretations. Using de Sadeleer's terminology, not all environmental principles are indeterminate rules or directing principles.[29] The principle of sustainable development[30] or common but differentiated responsibilities,[31] for example, appear as abstract concepts, ideas, and metaphors in international instruments and as part of the discursive practices of actors, including within organisations and international courts and tribunals. An example is Article 7 of the *Rio Declaration* itself which uses both concepts in their broad and abstract form. Article 7 provides that:

> States shall co-operate in a spirit of global partnership to conserve, protect and restore the health and integrity of the Earth's ecosystem. In view of the different contributions to global environmental degradation, states have *common but differentiated responsibilities*. The developed countries acknowledge the responsibility that they bear in the international pursuit of *sustainable development* in view of the pressures their societies place on the global environment and of the technologies and financial resources they command.[32]

This is not to suggest that these principles are never spelled out in any detail. They do not need to be epistemes that appear as abbreviated norms and can be heuristic devices with the potential to interplay with social processes involving groups. As abbreviations they are potentially more 'graspable, more memorable, and more easily applicable than the more numerous, more complicated legal rules'.[33]

---

[29] de Sadeleer, above n 4, 308–309. Sadeleer refers to the precautionary, preventive and polluter-pays principles as directing principles of environmental law.

[30] Marie-Claire Cordonier Segger and Ashfaq Khalfan, *Sustainable Development: Principles, Practices and Prospects* (2004); Daniel Magraw and Lisa Hawke, 'Sustainable Development' in Daniel Bodansky, Jutta Brunnée and Ellen Hey (eds), *The Oxford Handbook of International Environmental Law* (2007) 613.

[31] See Stone, above n 8; Edith Brown Weiss, 'Common but Differentiated Responsibilities in Perspective' (2002) 96 *American Society of International Law Proceedings* 366.

[32] Emphasis added.

[33] Michael Moore, *Educating Oneself in Public* (2000), 239.

As a general proposition, abstractness means that actors have a greater degree of choice in terms of how they can characterise a situation and the norm that applies to them and others in the particular circumstances. For instance, the term 'sustainable' is abstract because it does not specify the degree of harm that has to be avoided by the actor interpreting the term. Further, the prerogative to act sustainably may be interpreted differently depending on whether an actor is mining for uranium or considering whether to preserve a particularly sensitive area of the sea. In the latter instance, an actor's philosophical inclinations towards deep sea ecology might encourage them to do everything possible to protect a sensitive area of the sea because that is the most sustainable thing they can do for it.[34] In the former instance, however, anthropocentric views of consumption might shape how an actor sustainably mines for uranium.

The terms 'vague' and 'indeterminate' can also capture dimensions of the abstract nature of these environmental principles. For instance, the concept of 'differentiated treatment' in the principle of common but differentiated responsibility is both vague and abstract. It does not specify whether it applies in cases where multilateral agreements formally verbalise how actors are treated in a non-uniform way or if it also extends to how they are effectively treated.[35]

The abstract nature of environmental principles raises important issues as to whether in such abbreviated form they mean anything at all, and if they can be an episteme or a heuristic device. It is arguable that in an abbreviated form environmental principles are distinguishable from specific and precisely drafted norms, not just because they are usually stated in more abstract or vague terms but because they are potentially capable of 'instantiating', 'exemplifying' or 'illustrating' other norms such as rules. It is in this sense that reports prepared for the United Nations (UN) have tried to specify the norms that exemplify sustainable development.[36] Lowe

---

[34] For the variety of different ethical perspectives on environmental issues see, for instance, Joseph DesJardins, *Environmental Ethics: An Introduction to Environmental Philosophy* (4th ed, 2005).

[35] Stone, above n 8, 277. Stone at ibid has argued however that the literature on the principle has resolved this issue in favour of formally worded differentiated treatment.

[36] See, for instance, *Report of the Expert Group Meeting on Identification of Principles of International Law for Sustainable Development* UN Doc E/CN.17/1996/17/Add.1 (1996). The report lists, at para 3, 19 different 'basic principles and concepts of international law for sustainable development'. See also *Final Report of the Expert Group Workshop on International Environmental Law Aiming at Sustainable Development,* UN Doc UNEP/IEL/WS/3/# (1996).

makes a similar point when he says that sustainable development is no more than a name for a set of other norms.[37]

However, the argument against the determinacy of environmental principles as an episteme is put in a different way by Koskenniemi, who argues that they are often drafted with incompatible concepts and ideas built into them and actors are left to themselves to determine their meaning and normative impact at some future time.[38] His response is to suggest that, as a result of this indeterminacy, an environmental principle 'indicates the relevant values but leaves the determination of their normative impact to further process'.[39] The principle of sustainable development, which has been argued to 'mean different things to different constituencies', usefully highlights this point.[40] The International Law Association (ILA) has argued that a fundamental basis for its indeterminacy is that the principle prescribes the need to integrate social, environmental and economic growth concerns in decision making. Despite this, the ILA's 2006 report on 'International Law on Sustainable Development' argued that the indeterminacy of the norm can be dealt with by creating an integration model for the three pillars of sustainable development.[41] What this example suggests is that although environmental norms in their abstract

---

[37] Vaughan Lowe, 'Sustainable Development and Unsustainable Arguments' in Alan Boyle and David Freestone (eds), *International Law and Sustainable Development: Past Achievements and Future Challenges* (1999) 19, 30.

[38] Martti Koskenniemi, 'Peaceful Settlement of Environmental Disputes' (1991) 60 *Nordic Journal of International Law* 73, 76. He gives the example of Principle 21 in the *Declaration of the United Nations Conference on the Human Environment*, UN Doc A/CONF.48/14/Rev.1 (1972) ('*Stockholm Declaration'*) which gives states:

> the sovereign right to exploit their own resources pursuant to their own environmental policies, and the responsibility to ensure that activities within their jurisdiction or control do not cause damage to the environment of other States or of areas beyond the limits of national jurisdiction.

Koskenniemi, at ibid, points out that the principle states the values that underpin what states should do, but also confirms their sovereignty. He has expressed similar views in more recent publications: see Martti Koskenniemi, 'The Fate of Public international Law: Between Technique and Politics' (2007) 70 *Modern Law Review* 1; Martti Koskenniemi, 'International Legislation Today: Limits and Possibilities' (2005) 23 *Wisconsin International Law Journal* 61.

[39] Koskenniemi, 'Peaceful Settlement of Environmental Disputes', above n 38, 76.

[40] International Law Association, *Toronto Conference – International Law on Sustainable Development* (2006), [5], www.ila-hq.org/pdf/International%20Law%20on%20Sustainable%20Development/Report%202006.pdf.

[41] Ibid.

and abbreviated form can contain indeterminacies, they are not fatal in terms of the potential they have for creating change in the common and collective understanding of actors at the international level.

This kind of indeterminacy does not give certainty to actors seeking to apply environmental principles to situations that require direct application to particular problems but that is a different issue from whether they are capable of meaning anything at particular points in time. Kritsiotis, writing on self-defence in international law, takes a different approach to counter the potential indeterminacy of abstract norms. He argues that for a norm to reach a point of

> precision or common understanding . . . the rule might have to undergo considerable periods of development – of initiation and refinement, reception and reform, and then of further reception by states – before it can be said to gain its optimum legitimacy.[42]

This indicates that not all abstract norms 'instantiate', 'exemplify' or 'illustrate' other norms such as rules. Instead it points to the creative and extending potential they have because of the frame that they establish for actors to interact through them. In other words, some environmental principles might simply be epistemes or abbreviations couched in abstract terms without having any particular foundations in other norms or rules. The concept of 'interpretive communities' is a useful way of understanding how actors intersubjectively identify with each other in these cases by interacting within social contexts. Through their social practices, they create and sustain the meaning that is associated with norms.[43] In the case of environmental principles its 'interpretive community' is much larger than that of a rule that has a very defined context within which it can apply. An example of such an environmental principle is the principle of common concern.[44] Brunnée argues that there are a 'number of commonalities

---

42 Dino Kritsiotis, 'When states use armed force' in Christian Reus-Smit (ed), *The Politics of International Law* (2004) 45, 48 (citations omitted). Kritsiotis uses the term rules liberally to mean more abstract as well as closed norms.

43 Ian Johnstone, 'Treaty Interpretation: The Authority of Interpretive Communities' (1991) 12 *Michigan Journal of International Law* 371. Johnstone, drawing on the work of Stanley Fish who popularised the term 'interpretive community', has written at 381 that the 'interpretive task is to "uncover together" the meaning of the treaty; while auto-interpretation is carried on by individual participants, the process is essentially interactive'.

44 On the principle of common concern see also *Report of the Expert Group Meeting on Identification of Principles of International Law for Sustainable Development* UN Doc E/CN.17/1996/17/Add.1 (1996) paras 82–88. Internationally,

among common concern regimes', to identify the 'contours of a potential future customary framework'.[45] This suggests that the principle is likely to mean less in new social contexts than, for instance, sustainable development, which is more easily instantiated, exemplified, or illustratable because of the many situations in which it has been used in discursive and other kinds of practices. However, Brunnee's comments suggest that there is enough of an 'interpretive community' internationally to establish the necessary framework for understanding the principle of common concern. She even goes further and suggests that it has the potential to 'significantly widen the range of environmental protection obligations'.[46] The principle of common concern is an abstractly stated norm that is able to add creativity or to extend the relationship of groups of actors to the natural environment. It is a similar argument to the one by Kritsiotis, described above, because it suggests how abstract norms develop more epistemic certainty for actors through social practice.

These various arguments appear to point in one particular direction, which is to emphasise the importance of focusing on the social practices of actors to 'assess the *meaning*' as well as the role and function of environmental principles at particular points in time, rather than assuming that they are more or less determinate.[47] As abbreviated and abstract norms, environmental principles are still capable of generating an episteme, but one that depends on the nature of the social context within which someone is examining them. This is because as abstract norms principles can 'instantiate', 'exemplify' and 'illustrate' particular patterns of behaviour but they can also 'create' and 'extend' what they mean in new social contexts.

---

the principle was included in the preambles to the *United Nations Framework Convention on Climate Change,* opened for signature 9 May 1992, 1771 UNTS 165 (entered into force 21 March 1994) and the *Convention on Biological Diversity*, opened for signature 5 June 1992, 1760 UNTS 143 (entered into force 29 December 1993). The *Framework Convention on Climate Change* states in its preamble that 'change in the Earth's climate and its adverse effects are a common concern of humankind'. The *Convention on Biological Diversity*, also in its preamble, affirms that the 'conservation of biological diversity is a common concern of humankind'.

[45] Jutta Brunnée, 'Common Areas, Common Heritage, and Common Concern' in Daniel Bodansky, Jutta Brunnée and Ellen Hey (eds), *The Oxford Handbook of International Environmental Law* (2007) 550, 565–566.

[46] Ibid.

[47] Wiener, above n 12, 199.

3.4 CONCLUSION

This chapter has argued that environmental principles are open-textured because some of them are structured as norms which are open in terms of both the range of cases that they are relevant to and the normative solutions that they provide. Alternatively, environmental principles are often drafted as an abstract abbreviation which represents an episteme, the meaning and function of which are better specified in the context of social practice. This way of describing environmental principles does not appear to impact on their determinacy as norms, but instead emphasises their potential and depth of relevance for how actors create meaning from them in particular contexts. That is, instead of ignoring them as highly political, a focus on how they might constitute social processes or the participation of actors allows us to explore their relevance for learning and change in international law and politics.[48]

This chapter has not discussed the function of environmental principles, but instead has sought to define what they are in terms of their potential for common and collective social learning experiences of actors at the international level. Definitions of environmental principles and their function in society are inherently connected ideas. However, as was argued at the start of this chapter, the variability in terms of how environmental principles function is more easily explored with a description that facilitates this possibility. The next chapter critically engages with some scholars whose work discusses the relevance of open-textured or abstract norms for learning and change in international law and politics. Given that the focus of this work is on common and collective social learning it examines how scholarship has defined the possibility for it in the context of interactions that rely on abstract or open-textured norms.

---

[48] On this issue of the indeterminacy and political nature of principles see the special issue dedicated to an examination of Thomas Franck's work on *Fairness in International Law* (1995): (2002) 13(4) *European Journal of International Law* 902–1030.

4. Abstract and open-textured norms and social learning

4.1 INTRODUCTION

One of the arguments presented in the previous chapter was how the meaning of environmental principles emerges more clearly out of discursive and other practices of actors within a social context. The nature of an individual actor's interaction with others within a group of actors is important for understanding change. The importance of the interplay and constitutive relationship between abstract and open-textured norms and the participation of actors in international politics is often explicitly denied or not discussed in the literature on norms. This has the tendency, in practice, to narrow the relevance of norms like environmental principles at the international level. This is partly due to the dominant use of concepts such as obligation, consent, compliance, and enforcement in international law and politics, which seek to limit the participation of actors in defining and negotiating the meaning of the norm once its logic for action has been specified.[1] In such contexts, change through open-textured and abstract norms fails to fit traditional paradigms of regulation because it presumes more participation is necessary from actors for the norm to be meaningful in a particular context.

In other instances, where scholars have departed from using concepts like consent and obligation, they have not made the open-textured or abstract nature of norms an explicit part of their analysis. For instance, Franck, whose work on legitimacy and fairness has dominated discussions on norms in international law and politics, argues that an indicator of their legitimacy is their determinacy.[2] The determinacy of a norm, he argues, depends on how clearly any obligations are communicated and the extent to which they can be specifically applied to a situation.[3] The literary structure of a rule is an important indicator of determinacy, particularly in

1 Michael Barnett and Martha Finnemore, *Rules for the World: International Organizations in Global Politics* (2004).

2 Thomas Franck, *The Power of Legitimacy Among Nations* (1990) 90.

3 Ibid.

complex situations where a rule might appear simplistic. One can conclude from Franck's work on legitimacy that without a literary structure to enhance determinacy, a norm's potential impact on practice is likely to be minimal.[4] As a result of linking the determinacy of a norm with a tight sentence structure, Franck's work limits itself to a particular dimension of the relationship between the participation of actors within groups and it's relation to reifications in general.

In a very different body of work on the idea of legalisation, a group of scholars argue that the use and consequences of law in international politics depend heavily, amongst other factors, on the precision of rules.[5] According to these scholars, as a 'particular form of institutionalization' the role and effectiveness of international law depend on how precisely its norms are structured.[6] A theory that requires precision from norms gives no credence to the interplay and the co-constitution of participation and reifications. More particularly it assumes that open-textured norms give participants too much room to do what they want without considering that this could be positive in terms of the social learning that communities require to change (whether in a positive or negative way).

Three different approaches that work with the view that certain norms are open-textured or abstract are discussed in this chapter, especially where they pay attention to the participation of actors in social contexts and emphasise learning as the dynamic of change in international law and politics. In these three areas of discussion, open-textured and abstract norms are described as either a 'continuous social enterprise',[7] only progressively

---

4 This is naturally a simplification of what Thomas Franck has argued and how his work has been used by others. For instance, he has argued that a legitimate process will help reduce or resolve ambiguities inherent in the rule itself: ibid, chs 4 and 5; Thomas Franck, 'Legitimacy in the International System' (1988) 82 *American Journal of International Law* 705, 725.

5 Judith Goldstein, Miles Kahler, Robert Keohane and Anne-Marie Slaughter, 'Introduction: Legalization and World Politics' (2000) 54(3) *International Organization* 385. See also Judith Goldstein, Miles Kahler, Robert Keohane and Anne-Marie Slaughter (eds), *Legalization and World Politics* (2001).

6 Goldstein *et al*, 'Introduction: Legalization and World Politics', above n 5, 386.

7 See Stephen Toope, 'Emerging Patterns of Governance and International Law' in Michael Byers (ed), *The Role of Law in International Politics: Essays in International Relations and International Law* (2000) 91. See also Jutta Brunnée and Stephen Toope, 'International Law and Constructivism: Elements of an Interactional Theory of International Law' (2000) 39(1) *Columbia Journal of Transnational Law* 19; Jutta Brunnée and Stephen Toope, 'Environmental Security and Freshwater Resources: Ecosystem Regime Building' (1997) 91(1) *The American Journal of International Law* 26.

realisable,[8] or soft law.[9] In the previous chapter of this work, it was argued that environmental principles were open-textured or abstract norms. The chapter did not seek to define what role or function they might play at the international level, although it did engage with the idea that environmental principles as open-textured or abstract norms are potentially variable in terms of what they could do in social contexts.

This chapter reviews discussions in the more general field of international law and politics to assess the various ways that the role and function of abstract and open-textured norms are conceptualised in terms of framing or structuring social learning. It assesses approaches that describe the potential of common and collective learning at the international level through open-textured and abstract norms. The views described and examined here lend conceptual support to the possibility that actors can collectively identify with each other through norms, using frames as reference points for their interactions with each other. This discussion is important, given that chapter 2 described the idea of learning but did not establish its possibility in social contexts that develop out of interactions based around open-textured or abstract norms. It sharpens the particular analytical focus of this work by focusing on the importance of the interplay between norms and social processes as a way for groups to learn and for open-textured or abstract norms to establish themselves in concrete form amongst them.

4.2 INTERACTIONAL LAW, CONTEXTUALISED REGIMES AND EVOLUTIONARY APPROACHES TO NORMATIVITY

Toope and Brunnée have published a number of research papers, jointly and individually, where they develop a conception of law in the context of

---

[8] See, for instance, David Marcus, 'The Normative Development of Socioeconomic Rights Through Supranational Adjudication' (2006) 42 *Stanford Journal of International Law* 53; Audrey Chapman and Sage Russell, 'Introduction' in Audrey Chapman and Sage Russell (eds), *Core Obligations: Building a Framework for Economic, Social and Cultural Rights* (2002) 1.

[9] On the topic of soft law generally see, for instance, Christine Chinkin, 'Normative Development in the International Legal System' in Dinah Shelton (ed), *Commitment and Compliance: The Role of Non-Binding Norms in the International Legal Systems* (2000) 21; Ulrika Mörth (ed), *Soft Law in Governance and Regulation: An Interdisciplinary Analysis* (2004); Kenneth Abbott and Duncan Snidal, 'Hard and Soft Law in International Governance' (2000) 54(3) *Legalization and World Politics* 421.

constructivist thought in international relations.[10] They use terms such as interactional law, contextualised regimes, and evolutionary approaches to norms to describe how law is unique from a constructivist perspective on international law. They have also sought to describe a process, dominated by interaction amongst actors, whereby common understanding in specific regimes gradually evolves into law. On many occasions, writing jointly or individually, their work draws on environmental principles.[11]

Toope and Brunnée argue that the constructivist approach to international relations highlights the importance of common understanding amongst actors in particular contexts and how 'politics exists to help construct common understandings'.[12] This common understanding amongst actors is important because it provides a 'fruitful terrain' for the 'progressive elaboration of norms'.[13] Their engagement or interaction with each other or their participation in a common endeavour or common debate creates the potential for actors to develop 'shared meanings' about

---

[10] For instance see Toope, above n 7. See also Brunnée and Toope, 'International Law and Constructivism', above n 7; Brunnée and Toope, 'Environmental Security and Freshwater Resources', above n 7; Stephen Toope, 'Formality and Informality' in Daniel Bodansky, Jutta Brunnée and Ellen Hey (eds), *The Oxford Handbook of International Environmental Law* (2007) 107; Jutta Brunnée, 'Common Areas, Common Heritage, and Common Concern' in Daniel Bodansky, Jutta Brunnée and Ellen Hey (eds), *The Oxford Handbook of International Environmental Law* (2007) 550. Their work has been assessed (not necessarily as a collective body of work) by other scholars in the context of their own research. For instance see Christian Reus-Smit, 'Politics and International Legal Obligation' (2003) 9(4) *European Journal of International Relations* 591; Alhaji Marong 'From Rio to Johannesburg: Reflections on the Role of International Legal Norms in Sustainable Development' (2003) 16(1) *Georgetown International Environmental Law Review* 21; Jaye Ellis, *Soft Law As Topos: The Role of Principles of Soft Law in the Development of International Environmental Law* (DCL Thesis, McGill University, 2001); Owen McIntyre, 'The Emergence of an "Ecosystem Approach" to the Protection of International Watercourses under International Law' (2004) 13 *Review of European Community & International Environmental Law* 1; Jennifer Mitzen, 'Reading Habermas in Anarchy: Multilateral Diplomacy and Global Public Spheres' (2005) 99(3) *American Political Science Review* 401.

[11] See Brunnée and Toope, 'Environmental Security and Freshwater Resources', above n 7; Brunnée, above n 10.

[12] Toope, 'Emerging Patterns of Governance and International Law', above n 7, 102. Stephen Toope and Jutta Brunnée have co-authored many pieces together. Some of their individual works express their own views within the framework of constructivist approaches to international law. I use their various works to construct this part of this chapter.

[13] Ibid.

issues.[14] These shared meanings crystallise into norms,[15] making law or the formal expression of norms essentially rhetorical in nature or a 'form of practical reasoning'.[16] In a work focused on developing an interactional theory on international law, Brunnée and Toope argue that the patterns of interactions amongst actors give them the ability to 'learn' how they should read the 'social background against which any legal norm must be postulated and interpreted'.[17] The quality of the social background is, according to this view, an inherently important dimension of the potential normativity of rules. That is, the capacity of a norm to influence or have an effect depends on the interlocking beliefs or the mutual understandings that have shaped the background developments of particular contexts.[18]

The effect of these arguments is that norms evolve towards becoming legal in terms of what they mean to those within particular regimes. This is because they are based on common understandings or 'contexts of shared understanding' which change and also allow actors to make persuasive rhetorical claims from them.[19] Brunnée and Toope have argued that 'regimes are not static structures, but that they can evolve along a continuum from dialogue and sharing of information, to more defined frameworks for cooperation, to binding rules in a precise, legal sense'.[20] The common understanding which develops amongst actors serves as a 'frame' or 'starting points for argument' in terms of how a situation evolves.[21]

Toope has further argued that for legal norms to emerge and to play a role, they depend very much on how well the common understandings that have shaped them are constituted by values that are important for law.[22] These values can include, for example, justice and fairness.[23] This

---

14 Ibid, 102, 104–105. In a joint piece that Brunnée and Toope wrote together they emphasised this interactionist approach to law. They have relied on Lon Fuller to emphasise the idea that the role of law is to facilitate interaction: Brunnée and Toope, 'International Law and Constructivism', above n 7, 49.

15 Toope, 'Emerging Patterns of Governance and International Law', above n 7, 105.

16 Ibid, 102.

17 Brunnée and Toope, 'International Law and Constructivism', above n 7, 51.

18 Ibid, 52.

19 Ibid, 61.

20 Toope, 'Emerging Patterns of Governance and International Law', above n 7, 105.

21 Brunnée and Toope, 'International Law and Constructivism', above n 7, 61.

22 Toope, 'Emerging Patterns of Governance and International Law', above n 7, 102. In addition to values, Toope notes that two other variable are important to his consideration of whether norms evolve or not. The other variables are relative power and the interests of states and other actors.

23 Ibid.

is important because legal norms are about practical reasoning or rhetoric, and to function normatively they have to reflect the value inherent in law. An example given by Toope and Brunnée specifies how principles are likely to be effective because they encourage legitimacy by 'implementing fundamental legal values of equality, transparency, justice and fairness'.[24]

Their perspective on the emergence of law and its role as a rhetorical practice, or a norm that facilitates communication amongst actors, is instructive and important, in terms of how abstract and open-textured norms will function as frames.[25] Their focus on the evolution, amongst a group of actors, of commonly understood ideas towards rules of behaviour highlights how interaction can contribute to social learning. Their work is very much focused on differentiating law from politics, yet their conception of law is functionally designed to further communication amongst actors and to allow rhetorical arguments to be framed by it.[26] Social learning and the effectiveness or role of a norm like law, in other words, mature or develop as a result of the values inherent in a social process or mechanism rather than the interaction of actors about the subject matter of the norm itself. The idea of mutual engagement, the nature of the joint enterprise or a shared repertoire which could have arisen from the participation of actors in negotiating meaning with each other does not appear in itself valuable in term of the potential of norms to evolve beyond politics. Also, their consideration of the social practices of actors is limited to more formal and direct interaction which is contextualised within institutions. While this is the case in important instances, it is not the only way that the social practices of actors engage with or are intertwined with (and therefore reinforcing of) common understandings.

Toope has commented that 'regime-building is a fluid process, which proceeds along a continuum from mere coordination of viewpoints to

---

[24] Brunnée and Toope, 'Environmental Security and Freshwater Resources', above n 7, 58. The examples of principles they use are 'sustainable development, intergenerational equity, precaution, common concern, and the drainage basin' at 91.

[25] Brunnée and Toope, 'International Law and Constructivism', above n 7, 66. At 66, Brunnée and Toope, who adopt Lon Fuller's views to develop their approach to interactional law, point out that 'for Fuller, facilitating communication was the central purpose of law (rather than the promotion of any specific ethical agenda)'. It is in this respect that law 'in Fuller's description, can be viewed as even more important to the constructivist worldview than previously assumed, because law can help to create the conditions upon which changes of identity and interest rest': ibid 66.

[26] See ibid, 40.

the hardening of binding rules'.[27] This comment also creates a sense that what actors do fails to create the kind of intersubjectivity or interlocking beliefs that makes the ideas constitutive of what their preference will be. A common understanding of something sometimes does more than simply coordinate what actors do because of its constitutive effects on the preferences of a group of actors. It is more than coordination in a simple way because actors identify with each other. However, the ideas do not have to also be supervenient in the same way as capitalism or protecting the future generation might be.[28]

Most importantly, however, their conception of interaction has no way of identifying the variability and effectiveness of collective knowledge or culture in terms of their impact on the common understanding that actors develop amongst themselves in relation to norms. This is partly the reason why the legitimacy of a norm as a legal rule has to be found in values that are fundamentally reinforcing of law as a system of practice. Collective culture or knowledge-like thinking about the needs of the future generation can be supervenient on interaction in a way that reinforces how actors think about how they have developed their interlocking beliefs like norms. The association of norms with values inherent in legal forms of normativity underestimates the effect of the learning process through an episteme like open-textured or abstract norms. Likewise, their views do not appear to place value on the idea that the interplay of norms with social processes can in fact facilitate the normative pull that they expect from them, even in contexts where the values of law are not necessarily discussed or entertained as ideas.

4.3 SOFT LAW AND SOCIAL LEARNING

Many who write about principles, and environmental principles in particular, assume that they are soft law. Importantly, because soft law is often considered as abstract or open-textured, its function and role are considered in this section.[29] The idea of describing certain norms as soft law has become increasingly normal and is now seen as 'quite simply a

---

[27] Toope, 'Emerging Patterns of Governance and International Law', above n 7, 107.

[28] See chapter 2 and section 2.2.1 for discussions relating to collective knowledge or culture. Although more recently Brunnée, writing on her own, has highlighted how an abstract norm like 'common concern' has the potential to do more than engage actors in a simple act of cooperation: Brunnée, above n 10, 553–556.

[29] See, for instance, Friedrich Kratochwil, *Rules, Norms, and Decisions: On*

good thing, and no longer solely as the poorer cousin to hard international law'.[30] The challenge is that soft law as a concept can be interpreted as describing a wide and multifaceted role for norms in international politics. This is because it can serve as a rubric for describing a number of voluntary norms such as codes of conduct, recommendations and agreements more generally.[31]

Some scholars, however, associate abstractness or an open-textured nature with soft law, and use this as the sole means of distinction between it and hard law.[32] They argue that hard law is used as a tool of regulation to leave 'no legitimate choice to the rule follower'.[33] As a result soft law is also employed as a regulatory tool so that 'part of the decision to act in accordance with the rule is postponed and left to prospective rule followers'.[34]

Concepts like obligation and compliance are important for many scholars seeking to explain the effectiveness of soft law in regulating the collective behaviour of actors. For instance, Wellens and Borchardt refer to soft law as:

> the rules of conduct that find themselves on the legally non-binding level (in the sense of enforceable and sanctionable through international responsibility) but which according to the intention of its authors indeed do possess legal scope, which has to be further defined in each case. Such rules do not have in common a uniform standard of intensity as far as their legal scope is concerned, but they do have in common that they are directed at (intention of the authors)

---

the Conditions of Practical and Legal Reasoning in International Relations and Domestic Affairs (1991) 203.

[30] Jan Klabbers, 'Reflections on Soft International Law in a Privatized World' (2006) Erik Castrén Institute of International Law and Human Rights, www.helsinki.fi/eci/Publications/JKSoft_law_and_public.pdf. For critiques of soft law see Jan Klabbers, 'The Redundancy of Soft Law' (1996) 65(2) *Nordic Journal of International Law* 167; Jan Klabbers, 'The Undesirability of Soft Law' (1998) 67(4) *Nordic Journal of International Law* 381.

[31] Ulrika Mörth, 'Conclusions' in Ulrika Mörth (ed), *Soft Law in Governance and Regulation: An Interdisciplinary Analysis* (2004) 191, 198. On the topic of soft law generally see, for instance, Chinkin, above n 9; Mörth, above n 9; Abbott and Snidal, above n 9; Christine Chinkin, 'The Challenge of Soft Law: Development and Change in International Law' (1989) 38(4) *The International and Comparative Law Quarterly* 850.

[32] For instance, Goran Ahrne and Nils Brunsson, 'Soft Regulation from an Organizational Perspective' in Ulrika Mörth (ed), *Soft Law in Governance and Regulation: An Interdisciplinary Analysis* (2004) 171, 185–186.

[33] Ibid.

[34] Ibid.

and do have as effect (through international law), that the conduct of States, international organisations and individuals is influenced by these rules. . . .[35]

A different but simpler definition refers to soft law as 'rules of conduct which, in principle, have no legally binding force but which nevertheless may have practical effects'.[36] In both definitions there is a strong concern with defining soft law in terms of whether it is law or not, or whether it is binding or not, and the effect that this might have on international politics.[37] Without needing to discuss this perhaps unproductive impasse,[38] the literature on soft law has already suggested that this binary division between soft and hard law does not really explain its role in international affairs.[39]

In general terms, one can characterise soft law as to whether it functions as a 'transitional mode or a more independent form of regulation'.[40] A soft norm can become hard law depending on the processes in place to facilitate this.[41] This could take the form of the inclusion of soft norms into 'conditionality clauses' which are built into loan agreements by the World Bank or the International Monetary Fund. The continuous use of the same soft law encourages its gradual shift into contracts.[42] These kinds

---

[35] Karel Wellens and Gustaaf Borchardt, 'Soft Law in European Community Law' (1989) 14 *European Law Review* 267, 274.

[36] Francis Snyder, 'Soft Law and Institutional Practice in the European Community' in Steve Martin (ed), *The Construction of Europe – Essays in Honour of Emile Noel* (1994) 197, 198.

[37] For example, Jerzy Sztucki, 'Reflections on International "Soft Law"' in Lars Hjerner, Jan Ramberg, Ove Bring and Said Mahmoudi, *Festskrift Till Lars Hjerner: Studies in International Law* (1990) 549; Chinkin, above n 9.

[38] For a critique of the concept of obligation as the conceptual mechanism for dividing international law from politics see Reus-Smit, above n 10.

[39] Goldstein *et al*, 'Introduction: Legalization and World Politics', above n 5; Klabbers, 'Reflections on Soft International Law in a Privatized World', above n 30.

[40] Ulrika Mörth, above n 31, 191. The term regulation gives the impression that soft law is always deployed instrumentally to achieve particular effects. What is important for this discussion is the extent to which soft law, as potentially abstract and open-textured, draws on the dynamics of social learning to effect change as compared to how individual actors might feel obligated, compelled, or forced to do something.

[41] Mona Aldestam, 'Soft Law and the State Aid Policy Area' in Ulrika Mörth (ed), *Soft Law in Governance and Regulation: An Interdisciplinary Analysis* (2004) 11. See also Sia Spiliopoulou Åkermark, 'Soft Law and International Financial Institutions – Issues of Hard and Soft Law from a Lawyer's Perspective' in Ulrika Mörth (ed), *Soft Law in Governance and Regulation: An Interdisciplinary Analysis* (2004) 61.

[42] Aldestam, above n 41.

of descriptions of soft law say nothing about the nature of the norm or the dynamic of change that it helps engender but are more about the processes and social settings required to ensure that the transition happens.

There are also other subtle ways of describing how soft law functions in a transitional mode. Kratochwil, who sees soft law as being mostly of a higher degree of abstraction,[43] has written that it 'imposes an obligation' to 'seek a more specific and detailed solution to an issue without in itself imposing specific enforceable duties'.[44] From this perspective, soft law obligates actors to engage with each other to negotiate what a particular norm will mean when dealing with particular problems. That is, soft law expressly facilitates participation because any certainty in terms of meaning can only come from how the engagement of actors has been framed or generally structured rather than determined in particular ways.

This idea of soft law as transitional to hard rules has been put differently by other authors who argue that soft law is employed to create joint action in areas where there is no legal or political will to support particular decisions.[45] In this way a common ground for action is created or controversial issues are softened.[46] Soft law measures are seen as preparing 'the ground for hard law'.[47] It is capable of this because, as a non-binding abstract or open-textured norm, soft law gives plenty of room to actors with varying identities and preferences to engage with each other.

Both this view and the one above by Kratochwil describe collective action from two different perspectives. Kratochwil's emphasis on 'obligation' focuses his argument on the pressures created by the rule or context to move from a more general to a specific solution to the issues as they present themselves. That is, soft law generates dynamic whereby collective social learning happens because of the need or the obligation to solve problems. The alternative emphasises the potential for cooperation only to the point that actors might approach the need to take collective action because they realise that they have similar interests or perspectives.

Alternatively, Klabbers has noted that, in contrast to hard law which is

---

43 Kratochwil, above n 29, 203.

44 Ibid, 201. In a somewhat similar way Ulrika Mörth has also noted that soft law is potentially a 'transitional mode' of regulation: Mörth, above n 31, 191.

45 Kerstin Jacobsson, 'Between the Deliberation and Discipline: Soft Governance in EU Employment Policy' in Ulrika Mörth (ed), *Soft Law in Governance and Regulation: An Interdisciplinary Analysis* (2004) 81, 89.

46 Ibid.

47 Ibid.

used to coerce actors, soft law is used as a tool of persuasion.[48] In other words, persuasion works because actors have soft law at their disposal. This is different from the potential social influence that soft law can have when it is used to highlight what will be appropriate in a particular social context.[49] Either way, soft law functions to communicate information to individual actors through different dynamics of socialisation. The dynamics of socialisation referred to, explain the potential of soft law to influence individual rather than collective or communitarian behaviour.

As an independent form of regulation, soft law is also seen as a cognitive framework or a way to give actors a common language to use in particular contexts.[50] This means that soft law frames how actors come collectively to conceive of the reality of the particular situation that they are in with other actors.[51] In this way, concepts that develop or positions that actors take in relation to issues are set within the context of the common understanding amongst actors that is framed by soft law measures. Jacobsson points out that soft law does not have to ensure that there is total policy consensus amongst actors, but that it establishes a 'common nomenclature and the implicit cognitive framework for understanding the problems and functioning' of a particular situation.[52] She has noted that soft law on its own cannot achieve such effects without what she refers to as 'accompanying practices and institutional structures'.[53] In the context of her study of European Union employment policy, this means that 'collective action patterns' through meetings of states creates the dynamic whereby soft law becomes a part of the discourse of that collective experience.[54] Through regular meetings, for instance, states exchange 'policy knowledge and experience' and are socialised into particular ways of thinking about issues as framed by the cognitive structures set up by soft law.[55] Jacobsson's discussion uses the

---

48 Jan Klabbers, 'Reflections on Soft International Law in a Privatized World' (2005) 16 *Finnish Yearbook of International Law* 313, 314.

49 For the contrast between persuasion and social influence in relation to norms see Trine Flockhart, '"Complex Socialization": A Framework for the Study of State Socialization' (2006) 12(1) *European Journal of International Relations* 89.

50 See, for instance, Jacobsson, above n 45, 90; Kerstin Jacobsson, 'Soft Regulation and the Subtle Transformation of States: The Case of the EU Employment Policy' (2004) 14(4) *Journal of European Social Policy* 355.

51 See, for instance, Jacobsson, above n 45, 90; Jacobsson, above n 50; Martin Marcussen, 'OECD Governance through Soft law' in Ulrika Mörth (ed), *Soft Law in Governance and Regulation: An Interdisciplinary Analysis* (2004) 61.

52 Jacobsson, above n 45, 91.

53 Ibid, 98. On this see also Jacobsson, above n 50.

54 Jacobsson, above n 45, 94.

55 Ibid, 95.

idea of social practices as a way to bring peer pressure to bear upon states in conforming socially with the cognitive frame established through soft law measures.[56] Peer pressure is used as the dynamic that facilitates change, which is then framed by the soft law measures established for meetings.

The perspectives on soft law emphasise the importance of social practice in establishing the cognitive frame within the collective learning that goes on internationally. It focuses on the idea that the preferences and interests of actors emerge through interaction. In Jacobsson's own research the interaction of actors creates a situation whereby peer pressure influences the way states use soft law as cognitive frames for how they think and interact. What it seems to do is focus on common knowledge or discourse established through soft law as the basis for determining how an issue area can come to be governed. The focus on collective social learning, however, switches to what states do back in their respective countries. It does not emphasise the manner in which the common knowledge established through actors' interaction with each other is sustainable, constant, interlocking, or likely to reinforce a collective culture at the supranational level which can continue to structure and remain supervenient over what states continue to do. It relies on the potential transformative power of soft law at the actor level, and as a result does not adequately explain how it functions to engender collective social learning. This means that even though soft law creates meaning for actors, we are not any better off in terms of whether or not it is sustaining a culture that will remain after the peer pressure disappears.[57]

Importantly, Jacobsson's research points to participation through institutional structures as an important aspect of how soft law functions and the role it plays in the life of actors. However, it does not say much about the dynamics of social practice as part of collective social learning rather than individual learning. Therefore the question that it does not answer is whether or not the cognitive frame that soft law establishes contributes to social learning by impacting on more than the immediate interests of states. Importantly, it does not offer much in terms of how collective knowledge or culture shapes social practices, or the way soft law manages to serve as a frame in such diverse and plural environments of international relations.

This discussion has highlighted that soft law has a varied role in

---

56 See especially Jacobsson, above n 50.

57 On this see Rodger Payne, 'Persuasion, Frames and Norm Construction' (2001) 7 *European Journal of International Relations* 37. Against this see John Braithwaite and Peter Drahos who discuss the various material mechanisms available to place pressure on actors to do something: John Braithwaite and Peter Drahos, *Global Business Regulation* (2000), for example ch 22.

changing international relations and politics. Although both approaches described above highlight different ways that they can facilitate learning they also contain limitations for how abstract and open-textured norms might function to constitute the social practices of actors. This in turn limits how much we are able to observe more complex social learning. Soft law's characterisation as a cognitive framework works well in terms of how common knowledge might emerge from within organisations, but so far it has had a limited role in explaining how it might constitute the social processes in themselves. That is, peer pressure will be used to influence whether or not actors do something that is consistent with soft law, but little has been done that explores how soft law, as open-textured or abstract, constitutes peer pressure in a way that shapes what becomes common knowledge in the first place. The abstract qualities of soft law give actors a lot of room to cope with the peer pressure they might experience in a particular context. However, it also means that whether or not an actor exerts peer pressure will also be constituted by the norm. The difference here is that the level one is looking from determines whether or not one sees the variability in the way that abstract or open-textured norms can influence social practices. The cognitive frame in one instance seeks to regulate behaviour, but in the other situations it can do that and more (for instance, it might constrain, encourage, limit, create new opportunities) in terms of their relationship to the collective.

4.4 PROGRESSIVE REALISATION

The idea that with open-textured and abstract norms there is the expectation that actors will gradually define the details associated with them is captured in the idea of 'progressive realisation'. This idea for instance is encapsulated in Article 2.1 of the 1966 *International Covenant for Economic, Social and Cultural Rights*, which requires that states parties:

> take steps, individually and through international assistance and co-operation, especially economic and technical, to the maximum of its available resources, with a view to *achieving progressively the full realisation* of the rights recognized in the present Covenant by all appropriate means, including particularly the adoption of legislative measures.[58]

---

[58] *International Covenant on Economic, Social and Cultural Rights*, opened for signature 16 December 1966, 999 UNTS 3 (entered into force 3 January 1976) ('*ICESCR*'), Art 2.1 (emphasis added). It should be noted that this requirement is not found in the *International Covenant on Civil and Political Rights*, opened for

This provision was developed as part of the ICESCR to acknowledge that not all states would be able adequately to comply with the economic, social and cultural rights specified in the agreement.[59] The idea of progressive realisation assumes that the rights or norms are broad and that they allow for a variety of different interpretations depending on the actors involved. That is, the core content of the norm is difficult to identify. To measure performance against a norm is therefore likely to be problematic.[60] However, the role of rights or norms according to this idea is that actors can determine for themselves their level of participation or performance based on their realisation of what is domestically necessary.

The comment has been made that the requirement allowing for the progressive realisation of rights contributes to them becoming vague in the kind of obligation that they impose on states.[61] Others acknowledge that the idea of progressive realisation allows states gradually to comply with the variety of performance standards that are contained within each of the enumerated rights.[62] What is of most concern to those who have considered the implications of this idea is that, without a core minimum set of obligations in each of the enumerated norms that are abstract or open-textured, states cannot progressively realise the expectations of the drafters of the ICESCR.[63]

The idea that a norm can be progressively realised has not been applied outside the human rights literature, and even then its use within this field is still developing. The reason the idea of progressive realisation is discussed here is that it explicitly recognises the need for the state to participate in creating meaning for itself in terms of the relevance that rights have for states. It acknowledges that compliance, or the meaning of the rights for

---

signature 16 December 1966, 999 UNTS 171 (entered into force 23 March 1976). Another example includes the *Convention concerning Minimum Age for Admission to Employment*, opened for signature 26 June 1973, 1015 UNTS 297 (entered into force 19 June 1976); Art 1 requires that:

> [e]ach Member for which this Convention is in force undertakes to pursue a national policy designed to ensure the effective abolition of child labour and to raise progressively the minimum age for admission to employment or work to a level consistent with the fullest physical and mental development of young persons.

[59] See also Chapman and Russell, above n 8, 3.
[60] Ibid, 3–4.
[61] Marcus, above n 8.
[62] Chapman and Russell, above n 8, 4.
[63] See the various chapters in Audrey Chapman and Sage Russell (eds), *Core Obligations: Building a Framework for Economic, Social and Cultural Rights* (2002).

the state concerned, is not static and will develop over time. The literature discussing this idea, however, is concerned with how to measure performance against a right which a state can progressively realise for itself.[64] The idea of progressive realisation encourages and supports the active participation of the state in defining for itself how it will comply with a norm. It assumes that the rights serve as a frame which establishes the minimum set of obligations against which the participation of the state is measured at any particular point in time. The idea of progressive realisation has the potential to highlight whether and how states learn progressively to realise the manner in which they should comply with the norms in the ICESCR. Currently, however, the concern of the literature is with the nature of the obligations that the rights impose on states and how the idea of progressive realisation creates 'irreducible measurement complexities' for those trying to assess the performance of states.[65] The concept of progressive realisation usefully challenges states to continuously consider how they can improve their performance in relation to rights under the ICESCR. It does not, however, highlight the relationship that exists between states in terms of learning about the nature of those rights. It is ambivalent about the international community itself and the way that the social relations of states with each other will be conducive to learning about the rights.

4.5 CONCLUSION

In this work the role and function of principles have been conceptualised with social learning as the dynamic of change, rather than whether actors consent, feel obligated or have to comply with them. In this sense, the common theme amongst the different opinions canvassed in the discussions above is that open-textured or abstract norms can, as cognitive frames, have a variety of functional purposes depending on the particular approach discussed. In all three ways, actors learn through the norms to want something different or to develop collective interests which might be expressed as legal norms. The approaches to change described in this chapter can be illustratively summarised in two different ways. In one case (for example, the interactionist view of law and the transitional view of soft law) social learning is a common and collective experience because it is the community or actors as a collective unit within the institutional setting that decides what to do based on the cognitive frame established

---

64 See Chapman and Russell, above n 8.

65 Marcus, above n 8, 61, citing Chapman and Russell, above n 8, 5.

by the open-textured and abstract norms. The participation and interaction of actors with each other in this case is important because it creates common knowledge in a way that is framed by the principle, although not determined by it in a particularly strict way.[66]

In a different way (as an independent form of soft law regulation and concept of progressive realisation), an open-textured and abstract norm functions as a frame for the cognition of both the state seeking to influence and the state that is being pressured to think differently about its interests and behaviour. Collective social learning occurs only when a number of actors are equally influenced, persuaded or pressured to think within the boundaries of the frame established by the norm. So, a norm is commonly adopted when a significant number of international actors see it as their preferred way of doing something even though there is no formal agreement amongst them that they are bound by it.[67] In other words regulation and change more randomly affect the extent to which individual actors internalise norms and are constituted by them through their participation at the international level.

Whichever view one takes we are left without an adequate explanation of the way in which the social processes that define the nature of the participation of individual actors within a group interplay with or are constituted by abstract and open-textured norms. The difference between socialisation and social learning described in chapter 2 of this work is relevant here in terms of highlighting differences between the first and second approaches to change discussed above. Whereas the second approach described above reflects the potential influence of socialisation on how norms might create change internationally the first approach more closely resembles the approach to social learning. The difference is significant in terms of the potential for change in international law and politics, and also the depth of the cooperative environment that is established through norms. This is because, in the case of socialisation, actors can come together to develop norms, but once their individual interests disappear it will be difficult to

---

66 The following comments from two scholars highlight different aspects of the use of the term frame in relation to principle. Timothy Endicott states that '[a] norm, therefore, is not a frame, if by "frame" we mean a sharp boundary demarcating the discretion of a court': Timothy Endicott, *Vagueness in Law* (2000). Kratochwil makes a similar point in relation to rules by arguing that 'the lack of a "hard edge" to our concepts and the dependency of their meaning upon context defeats the argument that only the insistence on authoritative texts and their semantic can rescue us from the throes of uncertainty': Friedrich Kratochwil, 'How do Norms Matter?' in Michael Byers (ed), *The Role of Law in International Politics: Essays in International Relations and International Law* (2000) 35, 51.

67 See Braithwaite and Drahos, above n 57, ch 2.

explain why they might continue obeying the norms.[68] Also, it cannot explain the creative effect that norms can have on how actors respond to them because they are seen as directing or controlling behaviour.[69]

The alternative approach describes the potential for change in international law through collective social learning. However, this alternative appears to lack a theory about the nature and stages of learning within the change process in international law and politics. For instance, Toope and Brunnée discuss how the active participation of actors is important and their interaction with each other defines the background where social learning takes place. However, their description of interaction is not detailed enough to explain the nature of the transition from what is in the background to the emergence of norms in international law and politics. Although the link clearly exists in their work between what is socially learnt and what serves as the background to emerging legal regimes, there is no way of understanding how the variable nature of actors learning within groups contributes to the nature of change in international law and politics itself. This is necessary in the case of abstract and open-textured norms like environmental principles, the meaning and therefore function of which must depend on practice and which draw on intersubjectivity within groups where the learning has apparently taken place.

This chapter has illustrated the importance of social learning in the context of norms and change in international law and politics. Existing explanations of how this happens and why it is important confirm the main driver for this work and also the idea that learning is an important and necessary dynamic for change in international law and politics. In highlighting the difference between socialisation and collective social learning this chapter has sought to illustrate the difference in approach to common and collective change.

---

68 See generally the distinction in the approach to law from realism, rationalism, and social constructivism in Christian Reus-Smit, 'The politics of international law' in Christian Reus-Smit (ed), *The Politics of International Law* (2004) 14.

69 See, for instance, Martha Finnemore and Stephen Toope, 'Alternatives to "Legalization": Richer Views of Law and Politics' (2001) 55(3) *International Organization* 743.

5. The function and role of environmental principles

5.1 INTRODUCTION

Much of the scholarship on environmental principles is concerned with what various norms mean and their status in international law.[1] The aim of the discussion in this chapter is to review the scholarship which examines whether and how environmental principles change international law and politics. This is important for two reasons: it helps to support the argument that environmental principles have a role to play in changing international law and politics; and identifies the need to broaden the approaches to studying the nature of change in international law and politics and the potential role of environmental principles in that process. The discussion is structured around the realist and liberal traditions of what norms do in international relations.[2] This approach helps identify

---

[1] See Philippe Sands, *Principles of International Environmental Law* (2nd ed, 2003) 25–61, 231–290; Philippe Sands, 'International Law in the Field of Sustainable Development: Emerging Legal Principles' in Winfried Lang (ed), *Sustainable Development and International Law* (1995) 53; Ulrich Beyerlin, 'Different Types of Norms in International Law' in Daniel Bodansky, Jutta Brunnée and Ellen Hey (eds), *The Oxford Handbook of International Environmental Law* (2007) 425; Patricia Birnie, 'Twentieth-century marine conservation conventions adaptable to twenty-first century goals and principles?: Part I' (1997) 12(3) *International Journal of Marine and Coastal Law* 307, 311; Winfried Lang, 'UN-Principles and International Environmental Law' (1999) 3 *Max Planck United Nations Yearbook* 157.

[2] This distinction is commonly employed by social constructivists to highlight the utility of their own approaches to international relations and politics: see, especially, Christian Reus-Smit, 'The politics of international law' in Christian Reus-Smit (ed), *The Politics of International Law* (2004) 14. Liberal approaches to norms share much in common with moderate social constructivist approaches. The major difference is in the way that social constructivists prioritise how actors form interests during their interactions with others rather than prior to them. For this debate see Andrew Moravcsik, 'Bringing Constructivist Integration Theory Out Of The Clouds: Has It Landed Yet?' (2001) 2(2) *European Union Politics* 226; Jeffrey Checkel, 'The Constructivist Turn in International Relations Theory' (1998) 50 *World Politics* 324.

the limitations, when examining environmental principles, of relying on rational choice-based approaches to norms and their role in changing how actors within groups intersubjectively associate with each other. The discussion that follows categorises and contextualises the various scholarly approaches in terms of their relevant theoretical orientations.

5.2 MAXIMISING INTERESTS

Realist approaches view norms and principles as epiphenomenal to the reality of what actors, and states in particular, do internationally.[3] They exist only as far as the desires and interests of the powerful are being maintained and determined by these norms. The creation and use of environmental principles can be important modes of action for the powerful because of the potential they have to develop them in indeterminate ways or self-serving ways.[4] Actors essentially use principles to fulfil their desires and bring about certain desired effects within particular contexts.[5] To Goldsmith and Posner this means that the institution of international law 'emerges from states acting rationally to maximize their interests given their perceptions of the interests of other states and the distribution of state power'.[6] It also means that changing preferences will most likely create instability within a particular institutional context because of the way these norms can be manipulated.

---

3 See, for instance, Susan Strange, *States and Markets* (2003); John Mearsheimer, 'The False Promise of International Institutions' (1994–1995) 19 *International Security* 5; Joseph Grieco, 'Anarchy and the Limits of Cooperation: A Realist Critique of the Newest Liberal Institutionalism' (1988) 42(3) *International Organization* 485.

4 See John Braithwaite and Peter Drahos, *Global Business Regulation* (2000) 529.

5 Ibid. On the idea that norms and institutions ultimately serve the self-interest of actors see generally realist/neo-realist literature in international relations and neo-liberal adaptations to them. Neo-liberal theories in particular argue that actors cooperate to establish norms and institutions because the gains for them are likely to be more significant than if they sought to assert themselves based on their relative power. See in particular Robert Keohane, *After Hegemony: Cooperation and Discord in the World Political Economy* (1984); and Robert Keohane, *International Institutions and State Power* (1989).

6 Jack Goldsmith and Eric Posner, *The Limits of International Law* (2005) 8. See for instance, Ronald Mitchell, 'International Environment' in Walter Carlsnaes, Thomas Risse and Beth A Simmons (eds), *The Handbook of International Relations* (2002) 500, 504. Mitchell argues that if the United States had not applied pressure on Japan it would not have agreed with the moratorium on commercial whaling: ibid.

Given their abstract and open-textured qualities, principles are often found amidst regulatory contests. This means that the direction of a particular course of action can change through the contest of principles. To realists it also means that contests are dependent on, and often won, based on, the material power distributions of actors. It has been argued that actors may even shift or change forums to avoid having to be bound by particular principles and to be able to support others.[7] On this view environmental principles will support a range of initiatives, only in the sense that a range of actors will be able to define their interests through them and make moves against others to the extent that they can utilise their power effectively.

5.3 CREATING MORE EFFECTIVE AND EFFICIENT NEGOTIATION PROCESSES

Environmental principles can reduce transaction costs and uncertainty in the negotiation process that actors engage in within the international order.[8] Discussions of environmental principles in the context of negotiations define their role in instrumental terms. Environmental principles can help during negotiations because it is easier to use them rather than bringing a range of rules to the table.[9] Principles are seen as mechanisms that can help actors to develop faster, speedier responses to issues which might otherwise be extremely difficult to negotiate round.[10] It is easier to use them because they 'can serve different interests and goals'.[11]

When it is not possible to get states to consent to certain measures, principles can encourage negotiations and develop interests in particular general directions. In this way they are used rhetorically to shape arguments and views on particular issues.[12] They help actors with building

---

[7] See Braithwaite and Drahos, above n 4, 529.

[8] See Scott Barrett, *Environment and Statecraft: The Strategy of Environmental Treaty-Making* (2003) for an in-depth consideration of the neo-liberal approach to cooperation in international environmental law making.

[9] Braithwaite and Drahos, above n 4, 29.

[10] See, for instance, Lluís Paradell-Trius, 'Principles of International Environmental Law: an Overview' (2000) 9(2) *Review of European Community & International Environmental Law* 93.

[11] Braithwaite and Drahos, above n 4, 530. Another way to put this is to suggest that they help with consensus building: see David Hunter, James Salzman and Durwood Zaelke, *International Environmental Law and Policy* (1998) 273.

[12] See Braithwaite and Drahos, above n 4, 529; Frank Maes, 'Environmental Law Principles, their Nature and the Law of the Sea: A Challenge for Legislators'

'consensus or alliance around a particular principle or principles'.[13] In addition to being used by actors in instrumental ways, they can also serve as guidelines for the actions that states might take in relation to other international actors within the negotiating environment or the frameworks established afterwards.[14]

Also, the complexity of environmental issues makes them difficult to deal with in a specific way at the international level.[15] It is argued that environmental 'principles have a broader significance, and give rise to a wider variety of applications and execution than a rule of law, which defines duties and rights in a clearer way'.[16] In this way they apply themselves more generally to a wider variety of negotiations and potentially include within them more information from different sectors dealing with the protection of the environment. For instance, Braithwaite and Drahos have noted that 'the same principle can accommodate different goals. World's best practice can simultaneously serve the environmental goals of NGOs and the profit goals of firms'.[17] This provides actors in negotiations with a variety of opportunities to see their pre-existing interests integrated into agreements of solutions.

Principles can enrich how multilateral agreements are negotiated, especially if framework conventions have already come into force with abstract and open-textured norms within them.[18] Environmental framework conventions are often able to engender interest around principles and initially to create general approaches to problems or issues.[19] Another way to see this is that framework conventions identify 'common goals' or generate a 'cooperative perspective' that can drive further negotiations.[20] The determinate way in which rules work casts actors as in or out of particular situations, whereas principles can be seen to 'enrich' how actors react to each

---

in Maurice Sheridan and Luc Lavrysen (eds), *Environmental Law Principles in Practice* (2002) 59, 67.

13 Braithwaite and Drahos, above n 4, 530.

14 Maes, above n 12, 67.

15 Alexandre Kiss and Dinah Shelton, *International Environmental Law* (3rd ed, 2004) 204.

16 Maes, above n 12, 66, with a reference in the footnote to Hermann Mosler, *The International Society as a Legal Community* (1980) 73.

17 Braithwaite and Drahos, above n 4, 530.

18 See, for instance, Nicolas de Sadeleer, *Environmental Principles: From Political Slogans to Legal Rules* (2005) 269.

19 Paradell-Trius, above n 10; Kiss and Shelton, above n 15, 204.

20 Jutta Brunnée and Stephen Toope, 'Environmental Security and Freshwater Resources: Ecosystem Regime Building' (1997) 91(1) *The American Journal of International Law* 26, 42.

other. This means that negotiations following framework conventions are often seen as being framed and structured by the more open-textured norms contained in them. This is consistent with Bodansky's claim that principles embody commitments that are more generally defined and stated than rules.[21] In this way, principles shape the way in which more specific and determinate rules develop, as opposed to reflecting what a collection of them within a particular context or institution will look like in an abstract form.[22] The politics of a particular situation or context emerges out of abstract ideas rather than becoming encapsulated in that way through the connections between a series of rules.

Lastly, within the negotiation context, the openness or abstract nature of environmental principles allows a variety of other principles and rules to work together because they can accommodate each other in their application to a particular set of facts. Principles can withstand contradictions.[23] The interpreter of a situation can emphasise different principles to support what they want to achieve or pursue. It also means that in the case of disputes between actors a principle can co-exist in its operation with another principle. The framing of politics in the case of disputes between actors is likely to be very different from a situation with rules.[24]

5.4 ESTABLISHING FRAMEWORKS AND PARAMETERS FOR ONGOING CO-OPERATION AND INDIVIDUAL DECISIONS OF ACTORS

Aside from the role environmental principles play in actors negotiating agreements and their respective positions within them, they can do different things within institutions and regimes once they are codified or established customarily in some way. They can directly 'influence or even determine

---

21 Daniel Bodansky, 'The United Nations Framework Convention on Climate Change: A Commentary' (1993) 18 *Yale Journal of International Law* 451, 501.

22 Martti Koskenniemi, 'General principles: reflexions on constructivist thinking in international law' in Martti Koskenniemi (ed), *Sources of International Law* (2000) 359, 371.

23 de Sadeleer, above n 18, 307. See also Paradell-Trius, above n 10, 96.

24 See, however, Joseph Raz who argued with Ronald Dworkin about how different rules can apply to the one situation and thereby create similar disputes to principles because of the varying weight that they might have in a particular situation: see Joseph Raz, 'Legal Principles and the Limits of the Law' (1972) 81 *Yale Law Journal* 823.

the outcome of cases'.[25] In these circumstances environmental principles are seen as reducing uncertainty for actors in terms of what others expect from them and the kind of preferences that they can legitimately express and contest. In this way, environmental principles also reduce transaction costs for an actor, which encourages their entrenchment and continuous usage by them. The discussion in this part is broken up into the different ways in which environmental principles function to bring coherence to environmental law generally.

5.4.1 Framework and Parameters within which Actors Operate and the Creation of Rules is Directed

Environmental principles do not give very precise positive form to ideas, values or possibilities in relation to the expected behaviour that actors need to pursue within particular institutions or regimes. Even so, principles are seen to be able to create some positively stated framework or parameter for actors who have consented to them. Environmental principles can solve the ongoing co-operation problems for actors by giving them the means to assess their respective positions against other actors in the light of the parameters set by them. In this way they are seen as 'fostering coherence' in terms of how different states with very different needs and resources respond to issues relating to sustainability (which brings together social, economic and environmental concerns) and protection of the environment.[26]

In environmental law the piecemeal approach to defining problems has given rise to the need to work with principles that can give structure to policy and rules in a coherent way.[27] There is a greater need for 'guide-posts around which dispersed laws could be reassembled and structured within an entirely new rule-making entity'.[28] This has been put in a different way by certain scholars writing about international environmental law principles. For instance, Verschuuren suggests that '[p]rinciples go beyond concrete rules or policy goals; instead, they say something about

---

[25] See de Sadeleer, above n 18, 290–291.

[26] Marie-Claire Cordonier Segger, Ashfaq Khalfan, Markus Gehring and Michelle Toering, 'Prospects for Principles of International Sustainable Development Law after the WSSD: Common but Differentiated Responsibilities, Precaution and Participation' (2003) 12(1) *Review of European Community & International Environmental Law* 54, 60.

[27] See, for instance, United Nations Environment Programme, *Training Manual on International Environmental Law* (2006), [23]–[24] www.unep.org/law/PDF/law_training_Manual.pdf.

[28] de Sadeleer, above n 18, 258–259.

a group of rules or policies, they denote what a collection of rules has in common, or what the common goal is of a collection of rules (for instance a statute)'.[29] Others use metaphors, such as stating that principles are the 'backbone' for the rules of the international community and the 'cement that binds together the various and often disparate cogs and wheels of the normative framework of the community'.[30] In all these configurations, environmental principles reduce uncertainty for actors in the international community by developing a framework within which rules and the conduct of actors can develop. In these formulations, principles sit hierarchically above other rules because of the way they operate at a level of abstraction that is considered to be higher and more than what rules can do in practice. Within the legal institution they are considered to have a higher legal value than rules.[31]

Given the uniqueness of environmental law as providing relatively piecemeal solutions to protecting the ecosystem, it is argued by various scholars that environmental principles therefore 'serve as anchor-points for regulatory change and innovation'.[32] Environmental principles cause these rules or less abstract norms to develop, and in that way create the framework or serve as the backbone against which they operate. Environmental principles are seen to function in this way as norms that propel action in a particular direction as opposed to determining it.[33] Sadeleer argues that:

> principles should act as a first cause, a matrix from which more precise rules naturally follow. On that basis principles play an essential role in the construction of legal systems; reflecting values and guiding concepts, they transcend the rules of positive law and provide them with a rational structure. They thus represent one facet of a systematic process of rationalization which translates specifically into a logical systematization of the rules that make up the subject.[34]

---

[29] Jonathan Verschuuren, 'Sustainable Development and the Nature of Environmental Legal Principles' (2006) 1 *Potchefstroom Electronic Law Journal* 1 [15], www.puk.ac.za/opencms/export/PUK/html/fakulteite/regte/per/issues/ 2006 x1x_Verschuuren_art.pdf.

[30] Nicolas de Sadeleer, 'Environmental Principles, Modern and Post-modern Law' in Richard Macrory (ed), *Principles of European Environmental Law* (2004) 223, 226.

[31] Piet Gilhuis, 'Consequences of the Introduction of Environmental Law Principles in National Law' in Maurice Sheridan and Luc Lavrysen (eds), *Environmental Law Principles in Practice* (2002) 45. See also Jonathan Verschuuren, *Principles of Environmental Law: The Ideal of Sustainable Development and the Role of Principles of International, European, and National Environmental Law* (2003).

[32] Braithwaite and Drahos, above n 4, 571.

[33] For instance, Bodansky, above n 21; Braithwaite and Drahos, above n 4, 18.

[34] de Sadeleer, above n 18, 267 (citations omitted).

Sadeleer discusses how environmental taxation measures are essentially driven by the principles of prevention and polluter pays.[35] Although a direct fiscal measure, taxing consumers and producers for damage they cause to the environment is directly consistent with the principle that polluters must pay for their damage to the environment. The principle in this instance is not directly a concern for actors interpreting the fiscal measure, but it has naturalised its introduction and application.

Often, as common denominators for a range of actions the principles function well because of the broad minimum standards that they set. They give broad and abstract guidance to actors in terms of conduct that they see is legitimately expected of them within a particular institutional context or regime established by an agreement of some kind.[36] Various actors are also probably aware that others will move in the same direction as themselves within the particular regime. In this way, and as long as actors agree with the principles, they are seen as reducing uncertainty and make working within institutions a productive endeavour. Environmental principles in this way also reduce uncertainty because actors understand that if it becomes necessary courts might be able to resolve disputes based on the broad parameters set by the principles.

5.4.2 Common Denominator for What can be Legitimately Expected of Actors

Given the interdisciplinary and multiple frameworks that need to be included in environmental policy decisions by international institutions, principles are often seen as 'contributing to facilitating international environmental decision making of a less cumbersome and time-consuming nature'.[37] For instance, 'they allow it to proceed in an incremental manner and amidst disagreement and uncertainty'.[38] Within institutions they also encourage and assist negotiations between a variety of actors. In some institutions they create a networked relationship between businesses, NGOs, and governments by giving them a basis for negotiation and

---

35 Ibid, 265. See also Timothy O'Riordan (ed), *Ecotaxation* (1997).

36 On the idea of legitimate expectation in relation to environmental principles see Alhaji Marong, 'From Rio to Johannesburg: Reflections on the Role of International Legal Norms in Sustainable Development' (2003) 16(1) *Georgetown International Environmental Law Review* 21; Marie-Claire Cordonier Segger and Ashfaq Khalfan, *Sustainable Development: Principles, Practices and Prospects* (2004) 97.

37 Paradell-Trius, above n 10, 94.

38 Ibid, 96.

discussion.[39] Actors therefore know that the kind of discretion that international institutions can exercise will be developed consistently within the parameters established by the relevant principles. This creates less anxiety for actors about how they will fulfil their desires and interests. For states, within these institutions it also gives them the ability to reduce the transaction costs involved in managing non-governmental organisations and the interpretive communities that are established around them.

5.4.3 Give Meaning to and Define Specific Rules

In addition to applying directly to actors and situations, another significant role for environmental principles, is to help with the interpretation or practical formulation of rules. Principles can provide the 'basic ideas' for a variety of rules, and in that process also help with their interpretation or application.[40] In that way, they are seen as feeding ideas 'into the legal process' itself.[41] It is also argued that their abstract nature precludes them from applying to concrete situations.[42] For instance, Article 3 of the 1992 *United Nations Framework Convention on Climate Change* includes principles as a guide for how states should interpret the agreement.[43] It has been argued that to see principles simply as aiding interpretation is to underestimate their role in legal practice.[44]

5.4.4 Fundamental to Particular Institutions or Regimes

Some principles are considered to be fundamental doctrines of environmental law.[45] Not all principles, however, are potentially fundamental doctrines that frame political responses to protecting the environment. Some doctrines simply help to measure conduct and others can guide behaviour in a particular direction. Nonetheless, there are a variety of ways in which scholars describe their importance. For instance, Verschuuren has argued

---

39 See Verschuuren, above n 31, 26.

40 Gerd Winter, 'The legal nature of environmental principles in international, EU and exemplary national law' in Gerd Winter (ed), *Multilevel Governance of Global Environmental Change: Perspectives from Science, Sociology and the Law* (2006) 587, 592.

41 Ibid, 590.

42 Ibid, 595.

43 Opened for signature 9 May 1992, 1771 UNTS 165 (entered into force 21 March 1994) ('*Climate Change Convention*'). See Bodansky, above n 21, 501–502; Sands, above n 1, 231–232.

44 For instance, Verschuuren, above n 31, 13.

45 United Nations Environment Programme, above n 27, 23–24.

that environmental principles have a higher moral and/or legal value.[46] This idea is also shared by others who describe the principles of law as 'aspirational norms in the development and application of legal rules' that can 'serve as guidelines for the actions of states'.[47] These principles are seen as fundamental to a particular formal or informal institution or context, and in that way surpass in weight all other norms. In social interactions these abstract or open-textured norms are seen as shaping conversations because of their fundamental nature. For instance, the right to sovereignty is a fundamental principle of international law, and in relation to environmental protection an extension of this would be the prevention of serious harm to the territory of another state.

5.5 ACCOMMODATING TIME AND OTHER DISCIPLINES

It has been argued that the abstract and open-textured qualities of environmental principles give them greater capacity to remain relevant for actors with the passage of time.[48] Rules, because of their reliance on particular circumstances, could potentially lose their relevance with the passage of time and changing situations. Nicolas de Sadeleer comments on how principles are 'better suited to adapting to the shifting forms that characterize current public policy, including environment policy'.[49] Protection and preservation of the environment rely on many different disciplines and areas of the law.[50] Principles are also seen to give actors working with these norms more capacity to integrate the social and economic with the environmental issues in discussions and negotiations.[51]

These same arguments can also apply to scientific uncertainty that surrounds many environmental issues.[52] States would be reluctant to

---

[46] Verschuuren, above n 31, 19.

[47] Maes, above n 12, 67, referring in footnotes to Pieter van Dijk, 'Normative Force and Effectiveness of International Norms' (1987) 30 *German Yearbook of International Law* 9, 14–15.

[48] de Sadeleer, above n 18, 274; Kiss and Shelton, above n 15, 83.

[49] de Sadeleer, above n 18, 229.

[50] Joseph DesJardins, *Environmental Ethics: An Introduction to Environmental Philosophy* (4th ed, 2005) 8.

[51] See, for instance, Cordonier Segger et al, above n 26; Marie-Claire Cordonier Segger, Ashfaq Khalfan and Salim Nakhjavani, *Weaving the Rules for Our Common Future: Principles, Practice and Prospects for International Sustainable Development Law* (2002); Paradell-Trius, above n 10.

[52] Ibid, 94; Gilhuis, above n 31, 50.

make socio-economic decisions when there was scientific uncertainty in relation to pollution restrictions or changes in technology. Principles accommodate the uncertainty created by the lack of scientific knowledge, the passage of time, or the general interests of states in things other than regulation. This is because they do not specify how they should behave in particular ways. The reaction of states to particular norms can develop out of changes that occur in technology and science. As a result open norms more easily accommodate their needs based, for instance, on scientific developments.[53]

5.6 FUNCTIONAL ROLE IS TO SERVE INTERSTITIALLY

Lowe uses the idea of interstitiality to argue that environmental principles function not by changing the behaviour of actors in a particular way but 'by modifying the normative effect of other, primary norms of international law'.[54] The function of such interstitial norms is to 'direct the manner in which competing or conflicting norms that do have their normativity should interact in practice'.[55] These principles do not apply directly to regulate conduct or, as Lowe puts it, they have no independent 'normative charge of their own'.[56] Their importance comes from giving a tribunal the opportunity to exercise judgment in a coherent way when it comes to balancing competing primary norms.[57] He argues that these norms are essential to 'legal reasoning' proceeding in an uninhibited manner.[58] Their role in developing coherence comes from modifying the operation of competing primary norms. Its use, Lowe suggests, is not restricted to judicial tribunals but also to politicians, for example,[59] who

---

53 Ibid, 48.

54 Vaughan Lowe, 'The Politics of Law-Making: Are the Method and Character of Norm Creation Changing?' in Michael Byers (ed), *The Role of Law in International Politics: Essays in International Relations and International Law* (2000) 148, 213. See also Vaughan Lowe, 'Sustainable Development and Unsustainable Arguments' in Alan Boyle and David Freestone (eds), *International Law and Sustainable Development: Past Achievements and Future Challenges* (1999) 19.

55 Lowe, 'The Politics of Law-Making', above n 54, 216.

56 Ibid.

57 Ibid, 217.

58 Ibid.

59 Other circumstances suggested by Lowe include negotiating treaties, development loans, or environmental controls: ibid.

could have primary norms thrown at them which conflict with their own interests expressed in another primary norm.[60] These interstitial norms, like sustainable development, do not emerge through state practice or *opinio juris*, being the mechanisms for creating ordinary primary norms in international law.[61] Instead, they 'simply "emerge" from within the international legal system'.[62]

Lowe uses the idea of vagueness to differentiate between primary rules that are normative and interstitial norms that direct the way in which competing and conflicting primary norms will interact.[63] This is apparent from the way he characterises the 'polluter pays principle' as a primary norm. Also, he does not appear to list any other environmental norms in this category and uses examples such as reasonableness to further refine his arguments about the interstitial qualities of the principle of sustainable development itself. The main function or role of these interstitial norms is in establishing coherence within a system of primary norms. What Lowe does is discuss how a principle like sustainable development brings about coherence and develops the detail within the context of how characterisation of facts and legal issues is bound to create differences in terms of how actors approach primary norms. It also appears that Lowe tries to escape the consent driven nature of international law as he explains how interstitial norms originate. He argues that the basis upon which an abstract norm like sustainable development exercises its normative role is not based on consent. Although Verschuuren does not use the concept of interstitiality, he has also argued that principles serve as the intermediary between ideals and rules.[64] So the precautionary principle is the intermediary between rules and sustainable development, which is the ideal.

5.7 CRITIQUE AND CONCLUSION

This review illustrates that environmental principles play a variety of different roles and functions in international law and politics. Scholars have used the abstract and open-textured nature of environmental principles to describe the potential they have for strategic action.[65] This account

---

60 Ibid, 217.
61 Ibid, 216.
62 Ibid, 219.
63 Lowe, 'Sustainable Development and Unsustainable Arguments', above n 54, 34.
64 Verschuuren, above n 31, 20–31.
65 See, for instance, section 5.2.1. above.

suggests that they can be used as instruments to express an actor's self-interest in something. Given their abstract nature, actors are also able to mobilise them to contest with other actors who may use other principles to do the same.[66] Scholars writing generally on open-textured norms have argued that they play a communicative function or role, and as a result they are variable and multidimensional in terms of their social function. They argue that principles are seen in instrumental terms as directing or controlling something rather than as committed to 'responsiveness, dialogue, accountability and democratic participation'.[67] For instance, van Gestel and Evers, writing in a different field from environmental law, have suggested that the open-textured nature of norms can encourage creative responses by giving actors the choice to select from a variety of technological measures necessary to achieve a particular goal.[68] This alternative way of viewing the open-textured and abstract nature of principles means that an instrumental view takes a myopic perspective on how they function. It also means that when examining principles in the context of social processes within institutions, the strategic perspective on principles looks to how they are used to coerce others.

Alternatively, the scholarly literature argues more generally that environmental principles are objectives, goals, or ideals for particular regimes.[69] A comparison of the objectives and principles of a multilateral environmental agreement highlights the potential differences between them. For instance, Article 1 of the 1992 *Convention on Biological Diversity* sets out three very generally worded core objectives for its implementation.[70] They include: 'conservation of biological diversity, the sustainable use of its components and the fair and equitable sharing of the benefits arising out of the utilization of genetic resources'.[71] Additionally, Article 3 of the

---

66 Braithwaite and Drahos, above n 4, 530.

67 Willem Witteveen, 'Turning to communication in the study of legislation' in Nicolle Zeegers, Willem Witteveen, and Bart van Klink (eds), *Social and Symbolic Effects of Legislation Under the Rule of Law* (2005) 17, 41–42.

68 Rob van Gestel and Guido Evers, 'Communicative legislation: Can we trust certified management systems as tools for the interpretation and enforcement of environmental laws?' in Nicolle Zeegers, Willem Witteveen, and Bart van Klink (eds), *Social and Symbolic Effects of Legislation Under the Rule of Law* (2005) 183, 184.

69 See section 5.4.4 above.

70 Opened for signature 5 June 1992, 1760 UNTS 143 (entered into force 29 December 1993) ('*Biodiversity Convention*').

71 Ibid, Art 1. The *Vienna Convention on the Law of Treaties*, opened for signature 23 May 1969, 1155 UNTS 332 (entered into force 27 January 1980), Art 18

1992 *Biodiversity Convention* states two principles that have relevance for all who ratify the agreement:

> States have, in accordance with the Charter of the United Nations and the principles of international law, the sovereign right to exploit their own resources pursuant to their own environmental policies, and the responsibility to ensure that activities within their jurisdiction or control do not cause damage to the environment of other States or of areas beyond the limits of national jurisdiction.

The *Biodiversity Convention* intentionally separates its objectives and principles. Even though the principles stated in Article 3 are arguably abstract and open-textured, they are nonetheless more than just terms in the contractual relationship; they reflect a historically contingent conception of what it means to be a state within the international system.[72] Future agreements amongst actors, through resolutions, guidelines or protocols, build and expand on the established meaning of these principles. It is the principles' interplay with power that establishes their persuasive authority over the biodiversity regime, first by excluding any transnational actors that have no access to sovereign territory, and secondly by narrowing the field of states which can join the treaty regime. In other words the logic of what is appropriate as a state member is established in the way that the principle gives meaning to their identities in that context.[73] Further, and perhaps more importantly, the intersubjectivity, 'interpretative community' and 'lifeworld' accompanying the principles are more universally established and malleable even if they appear only in abstract and open-textured ways. The objectives instead rely only on the terms of the treaty itself for their meaning.[74] Objectives cannot function within the regime

requires that state parties to a convention or a treaty do not do anything that would defeat its 'object or purpose'.

[72] On this idea see particularly the work of Christian Reus-Smit, 'Politics and International Legal Obligation' (2003) 9(4) *European Journal of International Relations* 591, 593.

[73] This point should not be taken as suggesting that the principles apply independently of the agreement. For instance, during the negotiations of the *United Nations Framework Convention on Climate Change,* opened for signature 9 May 1992, 1771 UNTS 165 (entered into force 21 March 1994), the United States insisted that principles be included only if they could be interpreted in the context of the agreement: see Bodansky, above n 21, 501–502.

[74] On these terms see the discussions in chapter 3 of this book. This argument is not meant to suggest that the environmental principles apply independently of the agreement, but that the intersubjectivity behind them comes from outside the regime as well as the negotiations within it.

independently of those terms because of their limited capacity to work through social processes within the regime. Other than influencing the interpretation of particular terms within the treaty, they have limited utility for social learning within the biodiversity regime.[75]

As an extension of the idea that principles are objectives, goals or ideas, some arguments assert that environmental principles are fundamentally important to particular contexts or are 'fundamental legal norms'.[76] In other words, they have a dimension of 'weight' which, to some, differentiates the role of environmental principles from rules.[77] The difficulty with this proposition is that not all concepts have a core meaning and carry the same degree of depth in terms of what principles mean to actors within a particular institutional context. Concepts within principles can be vague or indeterminate in the sense that they do not drive a particular idea too far. The suggestion that an environmental principle is fundamental to how we protect the environment means that all actors identify with it and it has a similar degree of significance for them. As earlier chapters of this work have argued, the open-textured or abstract nature of environmental principles is important for the way they function. However, whether they are 'fundamental' to a particular institutional environment must surely depend on the manner in which actors engage with each other and the environmental principles as reification.

Finally, a majority of the scholarship on environmental principles views their function and role in terms of how they regulate the behaviour of states. This means that non-state actors are usually seen as part of an advocacy network that mobilises themselves as 'norm entrepreneurs',[78] or actors who use frames to give meaning to the way in which international debates are being conducted by states about particular issues.[79]

---

[75] Finnemore has argued that states adopt norms 'not as means to ends, but as ends in themselves – as affirmations of value about the kind of world people wanted and the kind of behaviour that was acceptable': Martha Finnemore, *National Interests in International Society* (1996) 129. This is another way of arguing that environmental principles as compared to objectives are not just ideas.

[76] The term is borrowed from Kiss and Shelton, above n 15, 203.

[77] Ronald Dworkin, *Taking Rights Seriously* (1977) 26; Beyerlin, above n 1.

[78] Ethan Nadelmann, 'Global Prohibition Regimes: The Evolution of Norms in International Society' (1990) 44 *International Organization* 479, 482.

[79] For instance, Margaret Keck and Kathryn Sikkink, *Activists Beyond Borders: Advocacy Networks in International Politics* (1998); Margaret Keck and Kathryn Sikkink, 'Transnational advocacy networks in international and regional politics' (1999) 51 *International Social Science Journal* 89; Richard Price, 'Reversing the Gun Sights: Transnational Civil Society Targets Land Mines' (1998) 52 *International Organization* 613.

Transnational actors like multinational corporations are not seen as subjects of environmental principles when they engage with others like them at the international level. There are a number of instances where transnational actors are collectively organised at the international level and introduced to environmental principles.[80] The blind spots of public international law to private actors and private authority are already a well-canvassed subject.[81] The review of the literature examining the role and function of environmental principles appears also to exclude the potential they have to contribute to social learning amongst other actors at the international level.

Whether scholars base their views on the abstract or open-textured nature of environmental principles or not, this chapter shows how environmental principles are significant for the role they play in terms of changing international law and politics. The narrow approach of the scholarship is in terms of how groups of actors learn, through their interactions with each other and environmental principles, collectively to develop new preferences and ideas about protecting the environment. In other words, the role and function of environmental principles for how actors develop collective interests or 'common interest patterns' need more work in the scholarly approach to this subject matter.[82] Also, because of the lack of attention given to intersubjectivity and existing 'life worlds' within institutions and groups, the literature does not fully capitalise on the potential of the abstract and open-textured qualities of principles. Concerns with institutionalising the role and function of environmental principles as legal, ethical, and moral, as compared to how groups socially learn, can discount their potential at the international

---

[80] The *OECD Guidelines for Multinational Enterprises* and also the *Global Compact* are important examples of how environmental principles identical to and essentially drawn from multilateral agreements developed by states are integrated within frameworks encouraging dialogue amongst transnational corporations. For the *OECD Guidelines for Multinational Enterprises* are part of the *OECD Declaration on International Investment and Multinational Enterprises*: Working Party on the OECD Guidelines for Multinational Enterprises, *The OECD Guidelines for Multinational Enterprises: Text, Commentary and Clarifications* (2001), www.olis.oecd.org/olis/2000doc.nsf/LinkTo/NT00002F06/$FILE/JT00115758.PDF.

[81] See, for instance, A Claire Cutler, *Private Power and Global Authority: Transnational Merchant Law in the Global Political Economy* (2003); Sol Picciotto and Ruth Mayne (eds), *Regulating International Business: Beyond Liberalization* (1999).

[82] Jutta Brunnée, 'Common Areas, Common Heritage, and Common Concern' in Daniel Bodansky, Jutta Brunnée and Ellen Hey (eds), *The Oxford Handbook of International Environmental Law* (2007) 550.

level. Confining environmental principles to a discussion of how they function legally can avoid highlighting the commitment of actors to 'innovative approaches to decision-making' or 'the adoption of environmental protection measures'.[83]

[83] Jaye Ellis, Soft Law As Topos: The Role of Principles of Soft Law in the Development of International Environmental law (DCL Thesis, McGill University, 2001) 160.

PART III

Environmental principles in practice

6. Power, environmental principles and the International Court of Justice

6.1 INTRODUCTION

Institutions like the International Court of Justice (ICJ) do not just solve disputes based on what the parties to a case have argued as the right and wrong interpretations of the law and the facts of the case.[1] Functionalist approaches to courts and dispute resolution bodies can appear disparagingly uninterested in the political character of their work.[2] This is partly symptomatic of the idea that institutions, especially dispute resolution bodies, are meant to apply their authority without bias or dispositions that favour anything outside their immediate jurisdiction.[3] Courts are meant to apply rules in a neutral and direct way and anything else is viewed as an inappropriate exercise of their power.[4] This is hardly the case and an abundance of research has shown it to be otherwise.[5] When the ICJ, for

---

1 For a general overview of the role of the ICJ in dispute resolution internationally see John Merrills, *International Dispute Settlement* (4th ed, 2006); John Collier and Vaughan Lowe, *The Settlement of Disputes in International Law: Institutions and Procedures* (1999) especially ch 7.

2 On a critique of the functionalist reaction to international organisations see Michael Barnett and Martha Finnemore, 'The power of liberal international organizations' in Michael Barnett and Raymond Duvall (eds), *Power in Global Governance* (2005) 161.

3 Ibid, 175. See also Cris Shore and Susan Wright, 'Policy: A New Field of Anthropology' in Cris Shore and Susan Wright (eds), *Anthropology of Social Policy: Critical Perspectives on Governance and Power* (1997) 3.

4 For instance, commentators criticised the ICJ for its pragmatic rather than strict legal reaction to the *Nuclear Tests Cases (New Zealand v France) (Interim Measures)* [1973] ICJ Rep 135, *(Merits)* [1974] ICJ Rep 457: R St J McDonald and B Hough, 'The Nuclear Tests Case Revisited' (1977) 20 *German Yearbook of International Law* 337; Thomas Franck, 'World Made Law: The Decision of the ICJ in the Nuclear Tests Cases' (1975) 69 *American Journal of International Law* 612; J J Ruiz, 'Mootness in International Adjudication: The Nuclear Tests Cases' (1977) 20 *German Yearbook of International Law* 358.

5 Generally see Michael Byers, *Custom, Power and the Power of Rules:*

instance, judges disputes and delivers its reasons for its decision it is also contributing more generally to change in international law and politics rather than applying in a neutral way a set of rules to the facts before it.[6]

This is visible in a number of ways. In the context of the work of the ICJ, Article 38(1)(d) of the *Statute of the International Court of Justice* states that judicial decisions are a 'subsidiary means for the determination of the rule of law'. Further, Article 59 also says that '[t]he decision of the Court has no binding force except between the parties and in respect of that particular case'. The combined effect of these two provisions is that judgments of the ICJ are important only for the parties to the dispute and should not be seen as playing a broader and more open-ended role in terms of change in international law and politics. This is, however, not an accurate impression of the influence that the ICJ has in a direct and diffused way with future state litigants and those who are judges of the Court themselves. There is a heavy reliance on previous opinions of the Court by those who litigate before it and in the way that judgments are constructed by the judges themselves.[7] It has been shown by Miller how other international courts and tribunals also commonly refer to

---

International Relations and Customary International Law (1999), especially 120–124; Andrew Coleman, 'The international Court of Justice and Highly Political Matters' (2003) 4 *Melbourne Journal of International Law* 29; Jared Wessel, 'Judicial Policy-Making at the International Criminal Court: An Institutional Guide to Analyzing International Adjudication' (2006) 44 *Columbia Journal of Transnational Law* 377; Tom Ginsburg, 'Bounded Discretion in International Judicial Lawmaking' (2005) 45 *Virginia Journal of International Law* 631. For an insightful review for instance of the way that the ICJ exercised its 'judicial-legislative function' in the area of the law relating to international watercourses in the *Case Concerning the Gabčíkovo-Nagymaros Project (Hungary/Slovakia)* [1997] ICJ Rep 7 see Eyal Benvenisti, 'Customary International Law as a Judicial Tool for Promoting Efficiency' in Eyal Benvenisti and Moshe Hirsch (eds), *The Impact of International Law on International Cooperation* (2004) 85.

6 For similar ideas in the context of other courts and tribunals see Duncan Kennedy, *A Critique of Adjudication (fin de siecle)* (1997); Joel Trachtman, 'The Domain of WTO Dispute Resolution' (1999) 40 *Harvard International Law Journal* 333. For a more direct study of the impact of the international dispute resolution bodies on international environmental governance see Tim Stephens, *International Courts and Environmental Protection* (2009).

7 See, for instance, Mohamed Shahabuddeen, *Precedent in the World Court* (1996), whose entire book is devoted to an exploration of the way in which judges are guided by their previous decisions and how litigants use them to frame how they argue before the ICJ. See also Robert Jennings, 'The Judiciary, International and National, and the Development of International Law' (1996) 45 *International and Comparative Law Quarterly* 1, especially 6–12.

decisions of the ICJ.[8] Additionally and importantly, Charney has shown how a variety of international courts and tribunals have developed similar coherent responses to a range of doctrinal issue areas in international law and politics including 'the law of treaties, sources of international law, state responsibility, compensation for injuries to aliens, exhaustion of domestic remedies, nationality, and international maritime boundary law'.[9] This is not to suggest however that international dispute resolution bodies are consistent in terms of how they contribute more generally to the governance of particular areas of international law.[10]

It has been argued that judgments of the ICJ can contribute to change not simply because they are 'legally binding'.[11] It is their persuasiveness that determines the extent to which states will refer to them or avoid them.[12] Jennings points to the 'practical authority and power' that is associated with certain judgments that determines their relative impact on states and as a precedent for the ICJ.[13] These ideas encapsulate the potential that international organisations like the ICJ have to exercise authority and power over 'knowledge claims' or the deference that states show to their claims to certain types and kinds of 'knowledge and truth'.[14] In other words, they highlight the potential role or function of the ICJ, for

---

8 Nathan Miller, 'An International Jurisprudence? The Operation of "Precedent" Across International Tribunals' (2002) 15 *Leiden Journal of International Law* 483. According to Miller at 489, by 2002 decisions of the ICJ had been referred to 111 times by other international courts and tribunals.

9 Jonathan Charney, 'The Impact on the International Legal System of the Growth of International Courts and Tribunals' (1999) 31 *New York University Journal of International Law and Politics* 697, 699. See also Jonathan Charney, 'Is International Law Threatened by Multiple International Tribunals?' (1998) 271 *Recueil des cours* 101.

10 See, for instance, Tim Stephens, *The Role of International Courts and Tribunals in International Environmental Law* (D Phil Thesis, University of Sydney, 2005) 7; and more generally Stephens, above n 6. Sands has argued that the contributions of Judge Christopher Weeramantry, as a judge of the ICJ, can be singled out in terms of the impact they have had on international environmental law: Philippe Sands, 'Pleadings and the Pursuit of International Law: Nuclear Tests II (*New Zealand v France*)' in Antony Anghie and Garry Sturgess (eds), *Legal Visions of the 21st Century: Essays in Honour of Judge Christopher Weeramantry* (1998) 601. This view highlights the significance of dissenting opinions of the ICJ as well as judgments of the Court itself.

11 For example, Byers, above n 5, 122.

12 Ibid.

13 Jennings, above n 7, 8.

14 Clark Miller, 'Democratization, International Knowledge Institutions, and Global Governance' (2007) 20 *Governance* 325, 331. See also Barnett and Finnemore, above n 2.

instance, to frame and structure the common and collective understanding of states which could litigate before them as well as the Court itself in terms of how it responds to issues at some future date.[15]

This chapter discusses the role and function of environmental principles in terms of how they interplay with or constitute power in the context of the ICJ developing and delivering its written judgments. The interaction of litigants with each other and the bench of the ICJ is important not just for conveying facts and positions on the law. It shapes how the Court itself learns what its views are on legal issues. The idea that the process of litigation can generate common knowledge which is shared amongst the judges of the bench and potentially states which may appear before them as well as the future make up of the ICJ is itself the kind of collective social learning that is the basis of the discussion in this chapter. It argues that the focus on social learning highlights the significant role that environment principles can play in decisions of the ICJ, notwithstanding their infrequent and direct use in the judgments of the Court in determining the legal position of the parties before them.

In section 6.2 which follows there is a description of the conceptions of power, as a diffuse persuasive force, which forms the basis of the discussion in the rest of this chapter. The concept of power is the social process used in this chapter to describe how learning takes place as part of the work that the ICJ does. This is followed in section 6.3 with a brief description of the use of environmental principles in the judgments of the ICJ. The variable nature of the function and role of environmental principles and their potential for social learning is examined in three ways in the sections that follow. Section 6.4 discusses how environmental principles interplay with the institutional power of the ICJ to frame the negotiations of states post-adjudication. Although it is arguable that this is a functionalist argument as referred to above, this section asserts that the background knowledge created during the dispute resolution process makes environmental principles more meaningful for the parties and shapes how they will use them.

Section 6.5 examines the productive power of ideas which are expressed

---

[15] There are different ways to assess the role and function of international courts and tribunals. For some interesting examples see, for instance, Robert Keohane, Andrew Moravcsik, and Anne-Marie Slaughter, 'Legalized Dispute Resolution: Interstate and Transnational' (2000) 54(3) *International Organization* 457; Wessel, above n 5; Wouter Werner, 'Securitization and Judicial Review: A Semiotic Perspective on the Relation Between the Security Council and International Judicial Bodies' (2001) 14 *International Journal for the Semiotics of Law* 345.

through the ICJ in dealing with legal issues before it or in generating a common understanding of an open-textured environmental principle. Sections 6.5.1 and 6.5.2 examine whether environmental principles as abbreviated abstractions refer to ideas that shape the Court's approach to the legal issues before it. They compare whether the precautionary principle and sustainable development had a diffuse role in producing the Court's approach in the *Case Concerning the Gabčíkovo-Nagymaros Project (Hungary/Slovakia)*.[16] This comparative study aims to assess whether environmental principles can shape how the ICJ develops a common understanding of an interpretive position in relation to issues before it. Section 6.6 discusses how the open-textured nature of Principle 2 of the *United Nations Declaration on Environment and Development*[17] or Principle 21 of the 1972 *Declaration of the United Nations Conference on the Human Environment*[18] has enabled the ICJ to generate a unique common understanding amongst several of its judgments which is a significant departure from its established understanding in international law. This chapter draws heavily from the *Danube Dam Case* because of the different ways in which environmental principles functioned in that case.

6.2 POWER AND SOCIAL LEARNING

There is no doubt that power is a ubiquitous feature of social and political life internationally. Our answers to how it is constituted enable us to better understand why actors at the international level behave the way they do towards each other and how they collectively form interests in particular matters. The concept of power is more widely interpreted and critiqued than probably any other idea used in the social sciences.[19] In international relations the realist conception of power still has a strong

---

[16] *Case Concerning the Gabčíkovo-Nagymaros Project (Hungary/Slovakia)* [1997] ICJ Rep 7 ('*Danube Dam Case*').

[17] UN Doc A/CONF.151/5/Rev.1 (1992) ('*Rio Declaration*').

[18] UN Doc A/CONF.48/14/Rev.1 (1972) ('*Stockholm Declaration*'). For differences between Principles 21 and 22 and the ICJ's formulation see Edith Brown Weiss, 'Opening the door to the environment and to future generations' in Laurence B De Chazournes and Philippe Sands (eds) *International Law, the International Court of Justice and Nuclear Weapons* (1999) 338.

[19] On varying arguments relating to conceptions of power see generally Stefano Guzzini, 'Structural Power: The Limits of Neorealist Power Analysis' (1993) 47(3) *International Organization* 443; Stefano Guzzini, 'The Use and Misuse of Power Analysis in International Theory' in Ronen Palan (ed), *Global Political Economy: Contemporary Theories* (2000) 53; David Baldwin, 'Power and

pull, although recent scholarship in the area has clearly sought to displace it.[20] However, there are different ways to conceptually categorise power which can in turn shape how we understand particular events and decisions. Varying conceptions of power can shape how we approach the work of formal or informal dispute resolution institutions at the international level and how environmental principles frame or structure change in international law and politics through them. This chapter adopts the conception of power discussed and theorised in international relations by Barnett and Duvall.[21] It is argued that their conception of power gives meaning to the kind of social process that enables learning to take place within the context of the work that the ICJ does in international law and politics. The international law literature makes use of the concept of power but draws upon international relations for its source of sociological inspiration.[22]

Barnett and Duvall define power as the 'production, in and through social relations, of effects that shape the capacities of actors to determine their own circumstances and fate'.[23] This is quite different for instance from the conception of power that sees it in relational terms. Baldwin, for instance, views power as the 'capacity to get somebody else to do what he or she would not have done otherwise'.[24] Their definition, Barnett and

---

International Relations' in Walter Carlsnaes, Thomas Risse and Beth A Simmons (eds), *The Handbook of International Relations* (2002) 179.

[20] See, for instance, Guzzini, 'Structural Power: The Limits of Neorealist Power Analysis', above n 19; Guzzini, 'The Use and Misuse of Power Analysis in International Theory', above n 19; Baldwin, above n 19; Joseph Nye, *Soft Power: The Means to Success in World Politics* (2004).

[21] Michael Barnett and Raymond Duvall, 'Power in Global Governance' in Michael Barnett and Raymond Duvall (eds), *Power in Global Governance* (2005) 1; Michael Barnett and Raymond Duvall, 'Power in International Politics' (2005) 59 *International Organization* 39. This chapter draws on Barnett and Duvall because of the particular way they have integrated moderate constructivist orientations into their conceptualisations of power. Also, the various authors whose work appears as part of the edited volume by Barnett and Duvall have adopted parts or all of their descriptions of the various ways to categorise power. For a review of the book by Barnett and Duvall see, for instance, Scott Solomon, 'Power and the Politics of Global Governance' (2006) 8 *International Studies Review* 327.

[22] The relationship between different conceptions of power in international relations and how they relate to international law is considered in, for instance, Richard Steinberg and Jonathan Zasloff, 'Power and International Law' (2006) 100(1) *The American Journal of International Law* 64; and in the context of customary international law see Byers, above n 5, esp ch 1.

[23] Barnett and Duvall, 'Power in Global Governance', above n 21, 8.

[24] David Baldwin, *Paradoxes of Power* (1989) 7, quoted in Guzzini, 'The

		Relational specificity	
		Direct	*Diffuse*
Power works through	*Interactions of specific actors*	Compulsory	Institutional
	Social relations of constitution	Structural	Productive

Source: Michael Barnett and Raymond Duvall, *Power in Global Governance* (2005), at 12.

Figure 6.1 Types of power

Duvall would argue, brings together two core dimensions which would differentiate it from Baldwin's. They characterise them as 'the *kinds* of social relations through which power works, and the *specificity* of the social relations through which power's effects are produced'.[25] They use their definition of power and the two dimensions of it to develop a taxonomy which avoids having to prioritise discussions of agency and structure.[26]

Barnett and Duvall's taxonomy of power is represented above in Figure 6.1. Their reference in Figure 6.1 to institutional and productive powers which this section describes express the nature of diffused power. Both expressions of power depend on actors interacting with each other within a social context to learn collectively rather than being directly required to adjust predetermined preferences. Additionally, as discussed in section 6.3 below, historically, the ICJ has not often used its direct power to adjudicate on the rights of parties in terms of their obligations to comply with

Use and Misuse of Power Analysis in International Theory', above n 19, 60. On David Baldwin's conceptions of power in international relations see David Baldwin, 'Neoliberalism, Neorealism, and World Politics', in David Baldwin (ed), *Neorealism and Neoliberalism: The Contemporary Debate* (1993) 3; David Baldwin, 'Force, Fungibility, and Influence', (1999) 8(4) *Security Studies* 173.

[25] Barnett and Duvall, 'Power in Global Governance', above n 21, 12.

[26] Ibid, 12–13. See in particular section 2.2.1 discussing how this book approaches the issue of agency.

environmental principles, especially in their abstract form described in chapter 3. The rest of this section discusses the ideas of institutional and productive power which are applied later in this chapter.

6.2.1 Institutional Power

Barnett and Duvall develop the concept of institutional power to refer to situations where actors indirectly control others through formal or informal institutions which mediate between international actors.[27] International institutions are examples of situations where states do not exercise direct control over each other but seek to influence the direction of change through their rules and procedures.[28] What is important for this conception of power is that the institution is not completely dominated by one actor. In fact, Barnett and Duvall draw from scholarship on the moderate versions of the social constructivist tradition in international relations, like Abbot and Snidal,[29] as well as more recently Barnett and Finnemore,[30] to argue that it is more likely for institutions to exist independently of 'specific resource-laden actors'.[31] This position is consistent with constructivist ontology which argues that power and interests do not exist independently of interactions within groups.[32]

The social distance between taking an action and the production of its effects on an actor is a feature of this particular conception of power. As a result of this 'lag between stimulus and effect', the exact effects or results of power cannot be easily identified.[33] This particular conception sees the potential of power to have effects even when the social relations between actors are at a 'physical, temporal and social distance' from each other.[34] This means that in institutions for instance, an actor cannot directly control or exercise power over someone else. Although ultimately the power of an actor can express itself in producing certain effects on others this will happen in a diffuse way. This is in contrast to the classical realist account of power which sees actor A as using their material,

---

[27] Ibid, 15.

[28] Ibid.

[29] Kenneth Abbott and Duncan Snidal, 'Why States Act Through Formal International Organizations' (1998) 42(1) *Journal of Conflict Resolution* 3.

[30] Michael Barnett and Martha Finnemore, *Rules for the World: International Organizations in Global Politics* (2004).

[31] Barnett and Duvall, 'Power in Global Governance', above n 21, 16.

[32] See, for instance, Steinberg and Zasloff, above n 22, 82.

[33] Barnett and Duvall, 'Power in Global Governance', above n 21, 11.

[34] Ibid, 12.

normative and symbolic resources directly to control actor B in a particular situation.[35]

Institutional power works through the 'socially extended, institutionally diffuse relations' that parties have with each other. This means that power will express itself only because actors socialise with each other through and within the structures established within the formal or informal institution. Barnett and Duvall's conception of institutional power is different from their conception of power as productive in terms of constituting particular social relations amongst actors. What institutional power shares with productive power is its focus on the diffuse way in which it can have effects on actors.

6.2.2 Productive Power

In contrast to the concept of institutional power, productive power is more concerned with constitutive social processes which are not necessarily controlled by 'specific actors' but which are effected through the 'meaningful practices of actors'.[36] The social processes that create meaning for actors in terms of what they can do in a particular context also shape their understanding of who they are and what they want. The idea of productive power, according to Barnett and Duvall, is a diffuse and general social process. This means that it is more about systems for considering the various ways in which actors interplay with reifications and meaning or episteme.[37]

In the context of environmental issues for instance, anthropocentricism is a system that signifies particular ways in which an actor will relate to others in terms of how they view their relationship to the environment. It is also a discourse or a 'social process and the systems of knowledge through which meaning is produced, fixed, lived, experienced, and transformed'.[38] In the context of anthropocentricism, it means that actors will socialise in a way that will shape intersubjective structures to favour human interests over the environment. As a result, the preferences of actors to protect nature for its own sake will be working against the deeper discourse that underpins the discussions being had about the environment. Drawing from Foucault, Adler and Bernstein have also argued in the context of international relations that knowledge is productive in the sense of defining and

---

35 Ibid, 13–15.
36 Ibid, 20–21.
37 Ibid, 20.
38 Ibid.

constituting the social reality or the order of global things.[39] Adler and Bernstein use the term knowledge to refer to the power that discourse has to produce particular social facts.

The idea of productive power is built on the conception of power as a 'social process' whereby actors come to see themselves as having particular social identities or capacities in particular situations.[40] The interaction of individual judges with each other and the litigants throughout the proceedings help produce the way they interplay with and are constituted by social facts, knowledge or reifications of various kinds, which include norms like environmental principles. This implies that this conception of power is inherently connected with the social learning process which shapes how the ICJ as a group of judges interpret particular facts or the law it is dealing with. In this sense power is expressed through the ICJ as opposed to something that it can wield against others.

6.3 THE INTERNATIONAL COURT OF JUSTICE AND ENVIRONMENTAL PRINCIPLES

Various international judicial, tribunal or arbitral bodies have jurisdiction and have dealt with aspects of international environmental law and in particular have employed environmental principles in their work.[41] However there is no single international court, tribunal or arbitral body that is designed to deal solely with international environmental law issues.[42] The focus of this chapter has been on the ICJ not because of the strength of its contribution to shaping international environmental law but its position as the most prominent dispute settlement body internationally which has a more general jurisdiction than courts and tribunals of issue-specific

---

39 Emanuel Adler and Steven Bernstein, 'Knowledge in power: the epistemic construction of global governance' in Michael Barnett and Raymond Duvall (eds), *Power in Global Governance* (2005) 294, 294–295. See also Virginia Walsh, *Global Institutions and Social Knowledge: Generating Knowledge at the Scripps Institution and the Inter-American Tropical Tuna Commission, 1900s–1990s* (2004).

40 Barnett and Duvall, 'Power in Global Governance', above n 21, 9.

41 For a list of these international courts and tribunals and their various contributions to the development of international environmental law see Stephens, above n 6.

42 For discussions relating to the need for an international court to deal with environmental matters see Amedeo Postiglione, *The Global Environmental Crisis: The Need for an International Court of the Environment* (1996); Alfred Rest, 'The Indispensability of an International Environmental Court' (1998) 7 *Review of European Community and International Environmental Law* 63.

regimes like the International Tribunal for the Law of the Sea or the World Trade Organization's Dispute Settlement Panel.[43] In 1993, the ICJ set up a permanent Chamber on Environmental Matters,[44] but it has yet to deal with a case that raises issues solely dealing with the protection of the environment or natural resources.[45]

It has been argued that dispute settlement has contributed very little to the development of international environmental law.[46] Although dispute resolution bodies have not used terms such as rules, principles or standards in a consistent way,[47] the scholarship often questions the functional value

---

[43] See Arts 34 and 36 of the *Statute of the International Court of Justice* establishing its general jurisdiction over international disputes. Weeramantry V-P, in his dissenting judgment in the *Request for an Examination of the Situation in Accordance with Paragraph 63 of the Court's Judgment of 20 December 1974 in the Nuclear Tests (New Zealand v France) Case* [1995] ICJ Rep 288, commented at 345 that the ICJ was 'situated at the apex of international tribunals' and as a result 'necessarily enjoys a position of special trust and responsibility in relation to the principles of environmental law, especially those relating to what is described in environmental law as the Global Commons'.

[44] International Court of Justice, *Communiqué 93/20 on the Establishment of a Permanent Chamber for Environmental Matters* (19 July 1993).

[45] The Chamber will not play a role in relation to environmental disputes until a particular party to a dispute refers a case to it: Stephen Schwebel, 'Ad Hoc Chambers of the International Court of Justice' (1987) 81 *American Journal of International Law* 831. On an unsuccessful attempt to refer cases to the Chamber see for instance the *Case Concerning the Gabčíkovo-Nagymaros Project (Hungary/Slovakia)*.

[46] Stephens, above n 10, 9. For studies that have explored the way in which dispute resolution bodies have used international environmental law see John Martin Gillroy, 'Adjudication Norms, Dispute Settlement Regimes and International Tribunals: The Status of "Environmental Sustainability" in International Jurisprudence' (2006) 42 *Stanford Journal of International Law* 1; Jeffrey Dunoff, 'Institutional Misfits: The GATT, the ICJ and Trade-Environment Disputes' (1994) 15 *Michigan Journal of International Law* 1043; Tim Stephens, 'Multiple International Courts and the "Fragmentation" of International Environmental Law' (2006) 25 *Australian Yearbook of International Law* 227; Stephens, above n 6.

[47] In 2005, for instance, the arbitral tribunal of the Permanent Court of Arbitration in its decision in the *Case concerning the Iron Rhine ('Ijzeren Rijn') Railway (Belgium v Netherlands)* stated that:

> [t]here is considerable debate as to what, within the field of environmental law, constitutes 'rules' or 'principles'; what is 'soft law'; and which environmental treaty law or principles have contributed to the development of customary international law . . . The mere invocation of such matters does not, of course provide the answer

(2005) Permanent Court of Arbitration [58]–[60], www.pca-cpa.org/upload/files/BE-NL%20Award%20240505.pdf.

of the abstract or open-textured nature of principles in directly resolving particular legal or social problems before them.[48] This means that the role and function of environmental principles in the adjudication of disputes gets classed and described as indeterminate or political in nature.[49] Perhaps the ethical and aesthetic rather than purely economic judgments that dominate environmental consideration make problems more protracted. That is, in environmental disputes it is not just the preferences of actors that are central to a conflict but deeper judgments about engagement with what is good or aesthetically valuable in terms of its conservation for the future generation.[50] The underlying issues in disputes cannot always be easily resolved before the courts by the application of narrow and predetermined rules. Alternatively abstract and open-textured norms challenge the potential legitimacy of the dispute resolution body were it to decide in favour of particular ethical or moral positions without appearing to base its judgment on the apparent consensual nature of international law and politics.[51]

The ICJ in the Advisory Opinion on the *Legality of the Threat or*

---

[48] For instance, Cesare Romano argues, referring to international environmental disputes, that '[a]djudication will be resorted to only if the law is fairly, but not too, clear or if the parties agree to give the dispute settlement body a large leeway or even to engage in creative law-making': Cesare Romano, 'International Dispute Settlement' in Daniel Bodansky, Jutta Brunnée and Ellen Hey (eds), *The Oxford Handbook of International Environmental Law* (2007) 1036, 1041. In relation to *jus cogens* norms see, for instance, Andreas L Paulus, 'Jus Cogens in a Time of Hegemony and Fragmentation – An Attempt at a Re-appraisal' (2005) 74(3–4) *Nordic Journal of International Law* 297. On the ineffectiveness of international courts and tribunals to make good use of broad environmental norms see, for instance, Tim Stephens, 'The Limits of International Adjudication in International Environmental Law: Another Perspective on the Southern Bluefin Tuna Case' (2004) 19(2) *The International Journal of Marine and Coastal Law* 177.

[49] Martti Koskenniemi, 'Peaceful Settlement of Environmental Disputes' (1991) 60 *Nordic Journal of International Law* 73.

[50] For an example of discussions relating to 'judgment' in environmental ethics see Mark Sagoff, *Price, Principle, and the Environment* (2004) ch 1. For a good discussion of environmental ethics and international environmental law see, for instance, Christopher Stone, 'Ethics and International Environmental Law' in Daniel Bodansky, Jutta Brunnée and Ellen Hey (eds) *The Oxford Handbook of International Environmental Law* (2007) 291.

[51] This does not seem to be uniquely an environmental law and politics issue. In the context of human rights for instance, Marcus has argued that there is a perception that socio-economic rights as abstract and open-textured norms are not justiciable: David Marcus, 'The Normative Development of Socioeconomic Rights Through Supranational Adjudication' (2006) 42 *Stanford Journal of International Law* 53.

Use of Nuclear Weapons[52] referred to its own version of what is known as Principle 21 of the *Stockholm Declaration* or Principle 2 of the *Rio Declaration*. This was not the first case in which the ICJ dealt with environmental issues,[53] but it was the best opportunity it had to directly apply environmental principles to a dispute before it.[54] Since its decision in the *Nuclear Weapons Advisory Opinion* the ICJ has dealt with many other cases that in some way concern the environment.[55] In the *Danube*

---

52 *Legality of the Threat or Use of Nuclear Weapons (Advisory Opinion)* [1996] ICJ Rep 226 ('*Nuclear Weapons Case*').

53 By 1996 the ICJ had decided the following cases that involved environmental issues: *Icelandic Fisheries Case* [1974] ICJ Rep 3; [1974] ICJ Rep 1975; *Nuclear Tests Cases* (*New Zealand* v *France*) (*Interim Measures*) [1973] ICJ Rep 135, (*Merits*) [1974] ICJ Rep 457; *Case Concerning Delimitation of the Maritime Boundary in the Gulf of Maine Area* (*Canada/United States of America*) [1984] ICJ Rep 246; *Nauru Case* [1992] ICJ Rep 240; *Case Concerning Maritime Delimitation in the Area Between Greenland and Jan Mayen* (*Denmark* v *Norway*) [1993] ICJ Rep 38; *Request for an Examination of the Situation in Accordance with Paragraph 63 of the Court's Judgment of 20 December 1974 in the Nuclear Tests* (*New Zealand* v *France*) *Case* [1995] ICJ Rep 288.

54 Cases before the ICJ that have concerned themselves somewhat indirectly with environmental harm include for instance the following that deal with the delimitation of maritime jurisdiction: *Case Concerning Maritime Delimitation in the Area Between Greenland and Jan Mayen (Denmark v Norway)* [1993] ICJ Rep 38; *Case Concerning the Land and Maritime Boundary Between Cameroon and Nigeria (Cameroon v Nigeria: Equatorial Guinea Intervening)* [2002] ICJ Rep 303; *Case Concerning Delimitation of the Maritime Boundary in the Gulf of Maine Area (Canada/United States of America)* [1984] ICJ Rep 246. On the relevance of these cases for environmental issues and sustainability in particular see Gillroy, above n 46, especially 20–22; Brown Weiss, above n 18. Also, see the following two cases before the Permanent Court of International Justice which concerned international watercourses: *Case Relating to the Territorial Jurisdiction of the International Commission of the River Oder (Czechoslovakia, Denmark, France, Germany, Great Britain, Sweden/Poland)* [1929] PCIJ (ser A) No 23; *Diversion of Water from the River Meuse (Netherlands v Belgium)* [1937] PCIJ (ser A/B) No 70. Potentially, the ICJ has had opportunities to deal with environmental harm, but for one reason or another cases before it did not go that far. For instance, in the *Case Concerning Certain Phosphate Lands in Nauru (Nauru v Australia)* [1992] ICJ Rep 240, the ICJ refused jurisdiction to Nauru but the facts of the case would have necessitated a consideration of environmental issues and law.

55 See, for instance, *Case Concerning the Gabcikovo-Nagymaros Project (Hungary v Slovakia)* [1997] ICJ Rep 7; *Fisheries Jurisdiction Case (Spain v Canada)* [1998] ICJ Rep 432; *Case Concerning Pulp Mills in the River Uruguay (Argentina v. Uruguay), Request for the Indication of Provisional Measures,* Order of 13 July 2006, 45 ILM 1025, and, more recently, Ecuador instituted proceedings against Columbia for the aerial spraying of toxic herbicides that is allegedly causing damage to people, animals, crops and the natural environment across

Dam Case the ICJ in its judgment referred to more than just its version of Principle 21 of the *Stockholm Declaration*. This is not to suggest that environmental principles have not been brought up by litigants in other cases before the ICJ or discussed as part of the dissenting opinions of the judges of the ICJ.[56] However, since the *Danube Dam Case* the ICJ has not in its formal and written deliberations which make up its judgment, used established and commonly referred to environmental principles.

The fact that this is the case points to the difficulty of directly applying environmental principles in the context of legal dispute settlement. Importantly, the fact that the ICJ has not commonly used environmental principles to exercise its compulsory power over litigants does not mean that principles cannot influence the social learning within the Court and for those states which might appear before it at some future time. The sections that follow aim to examine the variety of different roles and functions that environmental principles have and could play in relation to the work of the ICJ.

6.4 POST-ADJUDICATION DIALOGUE AND NEGOTIATION

Quite often, scholarship in international law views the judgments of international dispute resolution bodies as an end to the relevant social context that gives shape to the evolution of a particular dispute and its resolution.[57] In many instances negotiations continue beyond the actual

---

its border: see International Court of Justice, 'Ecuador institutes proceedings against Colombia with regard to a dispute concerning the alleged aerial spraying by Colombia of toxic herbicides over Ecuadorian territory' (Press Release, 1 April 2008).

[56] Judge Weeramanty has delivered dissenting opinions in a number of cases discussing the neglect of environmental issues and principles by the Court. For instance in the *Request for an Examination of the Situation in Accordance with Paragraph 63 of the Court's Judgment of 20 December 1974 in the Nuclear Tests (New Zealand v France) Case* [1995] ICJ Rep 288 he commented that the:

> International Court of Justice, situated at the apex of international tribunals, necessarily enjoys a position of special trust and responsibility in relation to the principles of environmental law especially those relating to what is described in environmental law as the Global Commons.

(Weeramantry V-P, dissenting at 345).

[57] An exception in the area of international environmental law is Cesare P Romano, *The Peaceful Settlement of International Environmental Disputes* (2000). In relation to the World Trade Organization see Gregory Shaffer, 'Power,

litigation before the international dispute resolution body. Given the social nature of interactions during the litigation process, post-conflict dialogue can continue in the shadow of the judicial decisions handed down by the institution dealing with an international dispute.[58] The institutional structure that sustains this post-conflict dialogue is the traditional diplomatic relationships that states have with each other. It is a collective effort because the ICJ's judgment is also a part of the engagement of parties with each other after the relevant judgment has been handed down by the Court. It is in this sense that the ICJ is arguably exercising power indirectly and diffusely. Given the 'social distance' between judging a case and what happens afterwards it is hard to see or to problematise the exact way in which the effects of the power of the ICJ can constitute the social learning between states.[59] Nonetheless, the Court frames social learning and the development of common knowledge amongst the actors by drawing on the open-textured nature of principles to communicate with them during their post-conflict negotiations.

Barnett and Duvall's conception of institutional power, as discussed earlier in this chapter,[60] requires that the institution be independent of the actors who operate within it. It is arguable that decisions of the ICJ create that sense of independence from the institution itself. That is, the post-conflict negotiations are conducted in the shadow of a judgment which has established certain normative implications for those discussions and created the necessary independence of the actors from each other. A resort to post-conflict negotiations conducted in the shadow of a judgment can give parties further reason to negotiate the real differences between them in terms of the ethical and aesthetic values that are at issue. Barnett and Finnemore have for instance argued that international organisations can frame issues in particular ways so that 'desired choices seem particularly compelling'.[61] That is, international organisations can 'structure situations and social understandings in ways that channel behaviour toward

---

governance, and the WTO: a comparative institutional approach' in Michael Barnett and Raymond Duvall (eds), *Power in Global Governance* (2005) 130.

58 See for instance the following two cases that continue for significant periods of time in the shadow of the decision of the ICJ: *Case Concerning the Gabčíkovo-Nagymaros Project (Hungary/Slovakia)* [1997] ICJ Rep 7 at paras 1–72; *Nuclear Tests Cases (New Zealand v France) (Merits)* [1974] ICJ Rep 457 at paras 253–455. On this issue see, for instance, Romano, above n 57, where he discusses both these cases in the context of the impact that the decision of the ICJ had on the dispute.

59 Barnett and Duvall, 'Power in Global Governance', above n 21, 11.

60 See section 6.2.1 above.

61 Barnett and Finnemore, above n 2, 176–177.

some outcomes rather than others'.[62] International organisations can engage in practices whereby they classify or invent social categories which can then constitute the way actors deal with the social world around them.[63] Adler and Bernstein, using a similar idea, argue that 'episteme' or intersubjectivity can be mobilised by institutions as a resource to shape what actors take to be valid knowledge or understanding of their social reality.[64]

Environmental principles can be mobilised by the ICJ to frame and structure post-conflict dialogue. The power that the ICJ has is further promoted in giving context to negotiations after the actual litigation has finished. What the litigating parties take as useful or valid for post-conflict negotiations is an expression of this power which would not be as easily mobilised without environmental principles. The importance of this point is that the litigating parties might not ordinarily use environmental principles to frame and structure their negotiations arising from a particular dispute. However, their interactions with each other as mediated through the ICJ give shape to their preferences in a way which can make the use of principles seem natural or the best way to continue dealing with the problems between them. It is in this sense that environmental principles can frame and structure ongoing discussions without giving the impression that actors are compelled to follow particular rules or narrow formulations of what is right or wrong to do. That is, the exercise of power using environmental principles is indirect because its effects are not going to be caused by the Court but will depend on how the parties interpret their positions in the light of the environmental principles that give meaning to how they view their circumstances.

The courts and tribunals, in doing so, interplay with environmental principles to harness their own institutional power in a 'responsive' or 'reflexive' way. The idea that international adjudication can facilitate litigants to resolve their disputes as opposed to ruling on all aspects of their case is not new.[65] Environmental principles give courts and tribunals a

---

[62] Ibid, 177. In their own work on international organisations, Barnett and Finnemore, above n 2, at 177–178 highlight how international organisations determine outcomes in the way they set agendas of meetings or classify practices of states.

[63] Ibid, 178. Barnett and Finnemore draw from the following text in making these points: Don Handelman, 'Commentary on Heyman' (1995) 36 *Current Anthropology* 280.

[64] Adler and Bernstein, above n 39, 294–298.

[65] In the context of the *Southern Bluefin Tuna Cases (New Zealand v Japan; Australia v Japan) (Provisional Measures)* (1999) 117 ILR 148, (1999) 38 ILM 1624 and *Southern Bluefin Tuna Case (Australia & New Zealand v Japan) (Jurisdiction*

chance to facilitate dialogue in a way that is potentially responsive to the needs of each party. In this way, they are central to the exercise of power by courts and tribunals by virtue of their abstraction and the communicative potential that comes with those qualities. However the power that the Court exercises is dependent on the knowledge that disputing states have about their situation from the decision of the case they are involved with. More importantly, the Court can be seen as creating opportunities for the parties to progressively realise for themselves the implications of environmental principles in the context of their disputes.[66] In this sense the environmental principles frame a state's ongoing self-realisation of its interests in the light of the dispute it is engaged in. This is obviously a different kind of argument from the power the ICJ has in terms of principles functioning as a justiciable standard of review of the conduct of parties.

In a number of instances the ICJ has engaged with or has relied on post-adjudicative negotiations to solve the real problem relating to a dispute. For instance, in the 1984 case on the *Delimitation of the Maritime Boundary in the Gulf of Maine Area (Canada v United States of America)*,[67] the ICJ had been asked by the United States to consider the ecological conditions of the marine ecosystem in the process of delimiting the Gulf of Maine as between it and Canada.[68] The proposal was aimed at ensuring that one of the parties to the dispute would have the entire region to maintain the integrity of the ecosystem for straddling stocks of fish. The particular stock was at that time being overfished and it was becoming endangered as a species. The ICJ decided the case by drawing the boundaries between the two countries across the natural ecosystem, and instead indicated that the United States and Canada should cooperate to take care of the environmental challenges caused by the Court's decision.[69] Gillroy argues that

---

and Admissibility) (2000) 119 ILR 508, (2000) 39 ILM 1359, see for instance: Douglas Johnston, 'Fishery Diplomacy and Science and the Judicial Function' (1999) 10 *Yearbook of International Environmental Law* 33; Bill Mansfield, 'Letter to the Editor: The Southern Bluefin Tuna Arbitration: Comments on Professor Barbara Kwiatkowska's Article' (2001) 16(2) *International Journal of Marine and Coastal Law* 361.

66 Marcus, above n 51. See also section 4.4. of this book for a discussion of this concept.

67 *Case Concerning Delimitation of the Maritime Boundary in the Gulf of Maine Area (Canada/United States of America)* [1984] ICJ Rep 246 ('*Gulf of Maine Case*').

68 *Gulf of Maine Case* [1984] ICJ Rep 246 at 278–288. On this case see also Brown Weiss, above n 18; Gillroy, above n 46, 21–22.

69 *Gulf of Maine Case* [1984] ICJ Rep 246 at 344.

the ICJ refused the sustainability argument put to it and instead opted 'for the more established sovereignty principle of "equity" to create the boundary, thus bifurcating the aqua-ecosystem between the two states'.[70] Although in this case the ICJ did not draw on an environmental principle to frame its post-conflict negotiations it explicitly relied on the fact that the United States and Canada would engage each other after the litigation to ensure that the ecosystem would not get harmed as a result of the decision by the Court.[71]

In the *Fisheries Jurisdiction Case (Germany v Iceland)*[72] the ICJ had to consider whether Iceland could legally extend its fisheries zone to 50 nautical miles and exclude fishing vessels from it. It decided that Iceland could not extend its fisheries zone beyond its territorial sea but that it had preferential rights over fisheries resources beyond it.[73] The ICJ also emphasised conservation measures in several statements, one of which is particularly important for this discussion. This is because of the way the Court connects to future interactions between the parties. It stated that:

> both Parties have the obligation to keep under review the fishery resources in the disputed waters and to examine together, in the light of scientific and other available information, the measures required for the conservation and development, and equitable exploitation, of those resources, taking into account any international agreement in force between them.[74]

In this paragraph the ICJ assisted the parties with their interactions beyond the dispute in that case by identifying matters that they needed to consider. It situated the decision to grant rights to Iceland within the language of having to negotiate in relation to conservation of the resources.

---

70 Gillroy, above n 46, 21.

71 Brown Weiss quotes the Court as telling the parties that the 'tradition of friendly and fruitful cooperation' between them would be used productively to prevent harm and ensure the conservation of the species: Brown Weiss, above n 18, 341 (referring to the *Gulf of Maine Case* [1984] ICJ Rep 246, [344]).

72 *Fisheries Jurisdiction Case (Germany v Iceland) (Merits)* [1974] ICJ Rep 175.

73 Ibid, [44]–[45].

74 Ibid, [64]. See also ibid, [37] in relation to comments by the Court as to the importance of conservation of fish stocks.

6.4.1 The *Danube Dam Case*, Post-adjudication, and Environmental Principles

The argument noted above will now be further illustrated through the work of the ICJ in the *Danube Dam Case*. This is because it is the most recent example of this kind of interaction between the ICJ and litigants after the formal dispute settlement process has finished, and also because the Court uses the language of environmental norms and principles to guide them. Also, Hungary and Slovakia, in the 1993 *Special Agreement between the Republic of Hungary and the Slovak Republic* giving the ICJ jurisdiction to hear the case, had agreed that the Court could also prescribe what the parties should do post-adjudication.[75] They had already agreed that they would further negotiate with each other once the Court had issued its ruling.[76] As a result, the ICJ judgment not only declared what the rights of the parties were in that case but also prescribed what they should do in their post-adjudication negotiations.

In the *Danube Dam Case* Hungary had argued that because of 'new requirements of international law for the protection of the environment' it was 'precluded' from performing the 1977 *Treaty Concerning the Construction and Operation of the Gabčíkovo-Nagymaros System of Locks*[77] with Czechoslovakia.[78] The 1997 Treaty had envisaged, amongst other things, that two locks would be constructed, one at Gabčíkovo (now Slovakia) and the other in Nagymaros (in Hungary). A dam would also be built at Dunakiliti which is in the Hungarian portion of

---

[75] See Art 2(1) of the *Special Agreement between the Republic of Hungary and the Slovak Republic for Submission to the International Court of Justice of the Differences between them concerning the Gabčíkovo-Nagymaros Project,* opened for signature 7 April 1993, (1993) 32 ILM 1293 (entered into force 28 June 1993).

[76] See ibid, Art 5.

[77] Opened for signature 16 September 1977, (1993) 32 ILM 1247 (entered into force 30 June 1978) ('1977 Treaty').

[78] *Case Concerning the Gabčíkovo-Nagymaros Project (Hungary/Slovakia)* [1997] ICJ Rep 7 at [111]. This description is not meant to be comprehensive, given that the arguments made in these sections do not need certain details from the case. For an excellent description of the facts and the processes leading to the judgment by the ICJ see Aaron Schwabach, *International Environmental Disputes: A Reference Handbook* (2005) especially 51–59. For discussions of the legal issues see Afshin A-Khavari and Donald Rothwell, 'The ICJ and the *Danube Dam Case*: A Missed Opportunity for International Environmental Law?' (1998) 22 *Melbourne University Law Review* 507; Constanze Schulte, *Compliance with Decisions of the International Court of Justice* (2004), ch 24; Romano, above n 57, 246–260; Philippe Sands, 'International Courts and the Application of the Concept of "Sustainable Development"' [1999] *Max Planck Yearbook of United Nations Law* 389.

the Danube. These projects would develop the energy and navigational needs of both countries as well as protect particular parts of the countries from flooding. When Hungary pulled out of the project one of its arguments related to its belief that further participation in the project would not be consistent with its obligations to protect the environment, water quality and fisheries of the Danube. Hungary argued that the project would be inconsistent with its international obligation to prevent environmental harm and the Slovak Republic's obligations to avoid serious environmental harm.[79]

Around 1991, Hungary showed little interest in further working on the project. Czechoslovakia (as it was in 1991) started constructing an overflow dam in Cunovo and later diverted the Danube through it. The effect of this diversion on riparian areas of the Danube in Hungary was considerable. This was in addition to reducing the total quantity of water that flowed through the Hungarian portion of the Danube. Before the ICJ, Hungary argued that the diversion of the Danube and the damage it had done were, amongst other matters, a breach of international environmental laws.[80] The specific points raised related for instance, amongst others, to the obligation not to cause damage to the environment beyond one's border, and the need to give prior notification to neighbouring states of activities that might cause significant damage to their environment.[81] The ICJ rejected both these arguments relating to the termination of the 1977 Treaty by pointing out that Articles 15, 19 and 20 allowed for these new and emerging concerns relating to the environment to be integrated into the bilateral contractual relationship.[82] The Court in turn argued that emerging environmental norms actually enhanced the relevance of Articles 15, 19 and 20 of that treaty for the parties as they negotiated their Joint Contractual Plan to implement the agreement.[83] The Court said that 'newly developed norms of environmental law' were 'relevant' to their ongoing discussions because they could use articles of the 1977 Treaty to agree on ways to integrate them into their negotiations.

The Court remained vague and ambiguous about what these 'newly

---

79 *Memorial of the Republic of Hungary in the Case Concerning the Gabčíkovo-Nagymaros Project (Hungary v Slovakia)* Vol 1 (2 May 1994) 198–203, www.icj-cij.org/docket/index.php?p1=3&p2=3&k=8d&case=92&code=hs&p3=1.

80 For Hungary's pleadings before the Court in relation to its allegations that Czechoslovakia had illegally diverted the Danube see ibid, 219–233.

81 Ibid, 198–203.

82 *Case Concerning the Gabčíkovo-Nagymaros Project (Hungary/Slovakia)* [1997] ICJ Rep 7 at [112].

83 Ibid.

developed norms of environmental law' were.[84] A study of the pleadings of Hungary in the *Danube Dam Case* shows that it drew heavily from environmental principles in developing its arguments before the Court. In particular, the precautionary principle as an expression of the more general principle of preventing harm to the environment gave shape to Hungary's arguments about terminating its obligations in relation to the 1977 Treaty.[85] In relation to the diversion of the Danube to operate the Gabčíkovo power plant, Hungary argued, amongst other things, that this was in breach of Czechoslovakia's (now Slovakia's) obligation not to cause damage to the environment beyond its sovereign borders.[86] These arguments clearly reflect those arising out of Principle 21 of the *Stockholm Declaration* and Principle 2 of the *Rio Declaration*. Additionally, Hungary also argued that Slovakia had an obligation to give it prior notification and to consult with it in relation to the construction of Variant C and the diversion of the Danube. This argument was based on, for example, Principle 19 of the *Rio Declaration*.[87]

What is most striking when one compares the arguments of Hungary with the final prescriptive judgment of the ICJ is the similarities in the kind of environmental principles it ended up referring to in its judgment to those Hungary used in its pleadings before the Court itself. Firstly, it should be noted that the ability of the Court to make such a recommendation was possible only because Articles 15, 19 and 20 were very generally worded to allow Hungary and Slovakia, in the process of drafting their future Joint Contractual Plans, to negotiate with each other in relation to the quality of the water of the Danube (Article 15), the protection of nature (Article 19) and fisheries (Article 20). For instance, Article 15 provided that the '[c]ontracting Parties shall ensure, by the means specified in the joint contractual plan, that the quality of the water in the Danube is

---

84 In the *Nuclear Tests case (New Zealand v France) (1995)* ICJ Rep 288, at para 64 the ICJ had stated that its order was 'without prejudice to the obligations of states to respect and protect the natural environment, obligations to which both New Zealand and France have in the present instance reaffirmed their commitment'. However the Court made no reference to what obligations it might be referring to. In discussing the *Nuclear Tests Cases (New Zealand v France) (merits)*, Sands, above n 10, 604–606, has argued that reference to the pleadings will give 'juridical assistance' to understanding what the Court might have had in mind.

85 *Memorial of the Republic of Hungary*, above n 79, 198–203.

86 Ibid, 219–223.

87 Ibid 223–228. Principle 19 provides that:

> [s]tates shall provide prior and timely notification and relevant information to potentially affected States on activities that may have a significant transboundary environmental adverse effect and shall consult with those States at an early stage and in good faith.

not impaired as a result of the construction and operation of the System of Locks'.[88]

This meant that in the context of doing what the 1977 Treaty between them provided, Hungary and Slovakia had to think about how new norms could be progressively applied to their particular situation. In this sense the Court's prescriptions were contextualised within the relationship that Hungary and Slovakia had already developed. Therefore Hungary's reference to environmental principles must naturally form the basis of any argument as to what those 'new norms' must be and in particular the way it must have shaped the manner in which the Court phrased its prescriptive formulations for Hungary and Slovakia in their post-adjudicative negotiations. The Court's judgment in this respect provided that it:

> is mindful that, in the field of environmental protection, vigilance and prevention are required on account of the often irreversible character of damage to the environment and of the limitations inherent in the very mechanism of reparation of this type of damage.[89]

This reference to the ideas contained within the precautionary principle or approach was further supported and built on when the Court also referred to sustainable development as a way forward for the parties to 'reconcile economic development with protection of the environment'.[90] One might argue that the exercise of institutional power to establish the precautionary principle and sustainable development (for instance) as a communicative norm in this particular context was paying lip service to the issue. However, from the Court's perspective it had found fault in the conduct of both parties in ways that encouraged them to negotiate further. It had also found that Slovakia had breached the 1979 Treaty and the law on watercourses by diverting the Danube into a variant C of the dam system that it had constructed to avoid economic loss. The application of the more determinate and closed rules to the case allowed the Court to engage both parties in a way that would have no impact on or would maintain its own legitimacy as a dispute resolution body.

Since the ICJ's judgment Slovakia and Hungary have not yet reached an agreement on implementation of the ruling.[91] The two countries were

---

88 Ibid, 184.

89 *Case Concerning the Gabčíkovo-Nagymaros Project (Hungary/Slovakia)* [1997] ICJ Rep 7, [140].

90 Ibid.

91 See United Nations Department for Economic and Social Affairs, Division for Sustainable Development, *International Rivers and Lakes Newsletter*

meant to have concluded an agreement in 2006 but this has not happened as yet. The effort made to move forward cooperatively reveals that the work of the ICJ was decisive in giving direction to the environmental concerns raised during the case. In 2001 a draft agreement between Hungary and Slovakia describing how they would proceed in their negotiations identified sustainable development and the principle of precaution as two guiding norms.[92] Additionally, minutes of 2006 meetings held between the two states reveal that they adopted a number of international agreements as between them to give further guidance in terms of their negotiations.[93] This is important, given that Slovakia during the litigation before the ICJ had emphasised the social and economic impacts of the diversion of the Danube rather than environmental concerns.[94]

What is instructive about the above discussion is that the power of the Court over the litigants in the post-adjudication phase was framed by emphasising that their 1977 Treaty could not be properly implemented in the future without reference to environmental norms. This was so even though the Court did not use environmental principles to declare whether the parties had acted properly or not in relation to the situation before them. This exercise of diffuse power over post-adjudicative negotiations was possible only because the Court defined the interests of the parties in particular ways during the litigation. For instance, the Court rejected

---

No 43 (2005) 3, www.un.org/esa/sustdev/sdissues/water/rivers_lakes_newsletter.htm; Beata Balogová, 'Slovakia again discussing completion of Gabčíkovo-Nagymaros', The Slovak Spectator' (Slovakia), 4 January 2006, www.spectator.sk/articles/view/2198.

[92] Preamble to the Draft Agreement Concluded between the Republic of Hungary and the Slovak Republic for the Purpose of Giving Effect to the Judgment of the International Court of Justice of 25 September 1997, www.gabcikovo.gov.sk/doc/prop/agreement-draft-02-04-01a.htm.

[93] Minutes of the Meeting of the Working Group on Legal Matters of the Delegations of the Government of the Slovak Republic and the Government of the Republic of Hungary on the Implementation of the Judgement of the International Court of Justice in the Case Concerning the Gabčíkovo–Nagymaros Project held in Budapest on 27 February 2006, www.gabcikovo.gov.sk/doc/prop/060227_zapEN.htm. Two examples include the adoption of the *Convention on Cooperation for the Protection and Sustainable Use of the Danube*, opened for signature 29 June 1994, (1996) 19 *International Environment Reports* 997 (entered into force 22 October 1998), and the *Convention on Wetlands of International Importance, Especially as Waterfowl Habitat*, opened for signature 2 February 1971, 996 UNTS 245 (entered into force 21 December 1975) ('*Ramsar Convention*'). These two agreements adopt important standards for the protection of the environment of fresh water and rivers.

[94] See in particular section 6.5.2 in this chapter for further discussion on the issue.

Hungary's arguments relating to the termination of the 1977 Treaty by pointing out that Articles 15, 19 and 20 provided the parties with opportunities to negotiate with each other about their environmental concern. This meant that the Court did not explicitly state its concerns relating to the environmental issues that Hungary had presented to it. However, were it not for the environmental principles, the Court might not have placed such emphasis on the dynamic nature of Articles 15, 19 and 20 of the 1977 Treaty. The meaning of such broad provisions within the 1977 Treaty was potentially established intersubjectively by the ICJ's directive that they need to be understood in the context of other environmental norms. This meant that the Court established the conditions for negotiating beyond the adjudication without stating what the outcome should be.

The open-textured nature of the norms allowed for the exercise of institutional power by the ICJ in ways that more determinate and defined rules would not have been able to achieve. This is because of the impact that this kind of decision would have on the legitimacy of the ICJ as an adjudicatory institution. However, the open-textured nature of the environmental principles contains a sufficient degree of intersubjectivity for the Court to use the contours of the norms for it to defer making policy decisions for the parties and instead create opportunities for them to give direction to how they will commonly identify with each other. In a way the ICJ is indirectly acknowledging the fact that the social or real interests of the parties should be resolved by opening up the legal domain in a way that ensures that this happens.[95]

6.5 PRODUCTIVE POWER AND ENVIRONMENTAL PRINCIPLES IN THE *DANUBE DAM CASE*

Gillroy has examined whether the principle of sustainable development is used by courts and tribunals as an adjudicatory norm.[96] That is, not as a norm capable of being used directly by parties to generate propositions about their position in relation to the other litigants but by the Court to resolve international disputes. This is partly consistent with the argument that Lowe made in a powerfully argued piece discussing the normative potential of sustainable development in ordering principles and rules in

---

95 Willem Witteveen, 'Turning to communication in the study of legislation' in Nicolle Zeegers, Willem Witteveen, and Bart van Klink (eds), *Social and Symbolic Effects of Legislation Under the Rule of Law* (2005) 17, 20.

96 Gillroy, above n 46.

international dispute resolution.[97] A different line of scholarship also argues that environmental principles such as sustainable development, for instance, facilitate intra-disciplinary integration amongst norms.[98] This argument, put forward initially by Boyle, views the role of environmental principles as reformulating '*existing bodies of law*' by integrating sustainable development into the development of other bodies of law such as, for instance, fisheries law or water law.[99] An alternative way to phrase this approach is to argue that the concept or idea of sustainable development can 'penetrate older terminology' especially when it is being applied some time in the future.[100] What these various discussions appear to draw from is the potential productive power of the idea of sustainable development in shaping the way courts, tribunals and litigants before them constitute the world that they are seeking to give meaning to and the reconciliation of different pulls on their preferences.[101] Unlike Lowe's, Gillroy's sustained

---

[97] Vaughan Lowe, 'Sustainable Development and Unsustainable Arguments' in Alan Boyle and David Freestone (eds) *International Law and Sustainable Development: Past Achievements and Future Challenges* (1999) 19.

[98] International Law Association, *Toronto Conference – International Law on Sustainable Development* (2006) 17, www.ila-hq.org/pdf/International%20Law%20on%20Sustainable%20Development/ Report%202006.pdf.

[99] Ibid (emphasis in original). See also International Law Association, *Report of the Seventy-First Conference – Berlin* (2004) 608–609.

[100] Alan Boyle, 'Relationship Between International Environmental Law and Other Branches of International Law' in Daniel Bodansky, Jutta Brunnée and Ellen Hey (eds) *The Oxford Handbook of International Environmental Law* (2007) 125, 131.

[101] Adler and Bernstein have used the word 'epistemes' to describe the discourses that have a productive effect on actors: Adler and Bernstein, above n 39, 295. The idea of 'episteme' in the context of the work of the ICJ has been raised in section 6.4. Episteme is 'the background intersubjective knowledge – collective understandings and discourses – that adopt the form of human dispositions and practices that human beings use to make sense of the world': ibid. This background knowledge, they argue, shapes institutions in particular ways. The application of this idea to international courts and tribunals is not new. For instance, Sandholtz and Sweet, in discussing the World Trade Organisation (WTO) dispute settlement regime, have argued that amongst other norms the most favoured nation principles generated a 'sophisticated, relatively autonomous domains of legal discourse, replete with their own stable or argumentation frameworks': Wayne Sandholtz and Alec Stone Sweet, 'Law, Politics, and International Governance' in Christian Reus-Smit (ed), *The Politics of International Law* (2004) 238, 254 (footnotes omitted). They have at 254 pointed out that by the 1980's an understanding of the most favoured nation principle 'could *only* be understood in light of the argument frameworks curated by the panels' of the General Agreement on Tariffs and Trade (emphasis in original). That is, the WTO dispute settlement regime produced the way that certain ideas about trade could be internationally understood.

and detailed examination of the decisions of the various international courts and tribunals shows that sustainable development is not an adjudicatory norm, even though he argues that it has a status as a principle of international law.[102]

Sustainable development however is inherently linked to the idea of protecting the future generation, but it does not specify whether anything in particular must be sustained for it. Sustainability, in other words, is not a particularly high standard to meet in terms of the minimum that is required to protect the interest of the future generation. As an abbreviated and abstract environmental principle it points to varying conceptions of how society and its relationship to nature can be constituted by different actors and groups. For instance, in a dispute about protecting an entire ecosystem some might argue that it is necessary to maintain its current condition because otherwise the future generation could be deprived of clean air. Others might equally argue that sustainability requires that the area also be preserved because it has aesthetic benefits as well.[103] A useful way of exploring whether a tribunal or court itself as an institution is constructed by the discourse or episteme that sustainable development or other environmental principles point to is to examine their reaction to issues before it in the context of whether they change established ideas or learn socially how to respond differently to qualitative tests in legal propositions before them.

In the following discussion in sections 6.5.1. and 6.5.2, the role of the precautionary principle and sustainable development are respectively analysed in the context of the *Danube Dam Case*, where the Court had to respond to qualitative tests which required a consideration of environmental issues. The question examined in each section is whether the common and collective understanding of the Court in interpreting the legal issues before it was produced by either environmental principle. Given the subtle nature of the social learning experience that is examined below the comparative study of the two principles is important for not drawing conclusions that cannot be confirmed by more than one example.

---

[102] Gillroy, above n 46, 2. He examines the concept in the work of the ICJ, the World Trade Organisation's Appellate Body, the International Tribunal for the Law of the Sea, and also the Human Rights Committee of the United Nations. He argues at 50 that the notion of sovereignty produces the responses of the ICJ not to favour sustainable development; and the WTO's Appellate Body defines itself by promoting economic efficiency which trumps sustainability as an adjudicatory norm.

[103] See Sagoff, above n 50, especially ch 2 for a discussion of sustainability and varying conceptions of it in the light of philosophical and ethical approaches to protecting the environment.

6.5.1 Ecological Necessity and the Precautionary Principle

In the *Danube Dam Case* the ICJ was asked by Hungary to consider whether its abandonment of the 1977 Treaty between itself and Czechoslovakia (at that time) was necessary because of the imminent peril to the ecological condition of the Danube, which made up an essential interest of Hungary.[104] This argument was based on Article 33 of the International Law Commission's (ILC) *Draft Articles on State Responsibility*,[105] which the ICJ recognised as reflecting customary law.[106] Article 33(1)(a) gives a state the ability to avoid its international law obligations if what it does 'was the only means of safeguarding' an 'essential interest' it has 'against a grave and imminent peril'.[107]

For this discussion, the key portion of Hungary's argument which caused difficulties was whether the damage caused by the diversion of the Danube by Slovakia was perilous and imminent.[108] The Court however examined whether the abandonment of the 1977 Treaty as a whole by Hungary could have been justified on the basis that the imminent peril of the entire project (and not just the Variant C that Slovakia had constructed in breach of the 1977 Treaty) necessitated the abandonment of the project. The Court therefore sought to interpret whether the potential harm inherent in the entire project established by the 1977 Treaty was imminent and perilous.

The Court's interpretation of the term 'imminent peril' in Article 33(1)(a) of the *Draft Articles on State Responsibility* and its application to the facts of the case highlight how the precautionary principle in this situation

---

104 *Case Concerning the Gabčíkovo-Nagymaros Project (Hungary/Slovakia)* [1997] ICJ Rep 7, [49]–[58]. See also *Memorial of the Republic of Hungary,* above n 79, 283–298. On this issue and how it relates to the precautionary principle see Daniel Dobos, 'The Necessity of Precaution: The Future of Ecological Necessity and the Precautionary Principle' (2001) 17 *Fordham Environmental Law Journal* 375.

105 (1980) 2(2) *Yearbook of the International Law Commission* 20 ('*Draft Articles on State Responsibility*'). On the concept of ecological necessity in the work of the ILC see James Crawford, *The International Law Commission's Articles on State Responsibility* (2002) especially 180–182.

106 *Case Concerning the Gabčíkovo-Nagymaros Project (Hungary/Slovakia)* [1997] ICJ Rep 7, [51].

107 International Law Commission, *Draft Articles on State Responsibility*, August 4, 1998, 37 ILM 44. See also the *Report of the International Law Commission on the Work of its Thirty-Second Session*, UN Doc A/CN.4/318/ADD 49 (1980).

108 *Memorial of the Republic of Hungary,* above n 79, 289–290.

failed to frame its approach. The Court said the following in relation to these words:

> The word 'peril' certainly evokes the idea of 'risk'; that is precisely what distinguishes 'peril' from material damage. But a state of necessity could not exist without a 'peril' duly established at the relevant point in time; the mere apprehension of a possible 'peril' could not suffice in that respect. It could moreover hardly be otherwise, when the 'peril' constituting the state of necessity has at the same time to be 'grave' and 'imminent'. 'Imminence' is synonymous with 'immediacy' or 'proximity' and goes far beyond the concept of 'possibility'.[109]

This view of Article 33(1)(a) presumes that a peril must be unavoidable at the time it is said to eventuate if a state is allowed to evoke the state of necessity argument.[110] The Court went on to find that the perils claimed by Hungary were 'uncertain' in 1989 when Hungary took steps to terminate its obligations under the 1977 Treaty.[111] It appears to have been looking for scientific 'certainty' in terms of whether there was likely to be some kind of peril for the essential interests of the state of Hungary. The Court was certainly not bothered by the argument that the potential peril, if ascertained with certainty, might be in the future because it stated that:

> a 'peril' appearing in the long term might be held to be 'imminent' as soon as it is established, at the relevant point in time, that the realization of that peril, however far off it might be, is not thereby any less certain and inevitable.[112]

The key reason therefore for rejecting Hungary's argument appears to have been its assessment of the 'uncertainty' of whether the project was perilous enough to warrant terminating it.

As stated above, the Court did acknowledge that the term 'peril' evoked the idea that the material damage had not actually occurred but that there was a 'risk' that it might happen. The ILC in its commentary on the *Draft Articles on State Responsibility* does not appear to have said anything about the level of scientific certainty required to assess the nature of a 'peril'.[113] Therefore the Court in this case had the potential to determine for itself whether it would make a policy choice to adopt the idea that the

---

109 *Case Concerning the Gabčíkovo-Nagymaros Project (Hungary/Slovakia)* [1997] ICJ Rep 7, [54].

110 See Erika Preiss, 'The International Obligation to Conduct an Environmental Impact Assessment: The ICJ Case Concerning the Gabčíkovo-Nagymaros Project' (1999) 7 *New York University Environmental Law Journal* 307.

111 *Case Concerning the Gabčíkovo-Nagymaros Project (Hungary/Slovakia)* [1997] ICJ Rep 7, [57].

112 Ibid, [54].

113 On this see also Dobos, above n 104.

reference to the level of risk in the term 'peril' had to be established with certainty or could be left uncertain.[114] It has been argued that the precautionary principle refers to 'uncertain risk' rather than requiring states to take preventative action when risk has been established with certainty.[115] Hungary in its written memorial to the ICJ in the *Danube Dam Case* referred to the precautionary principle as an expression of the concept of prevention in international environmental law. In this way it also highlighted the idea that risk could also be uncertain.[116]

The difference between certain and uncertain risk is in the 'link of cause and effect between an event that might occur and the damage anticipated as a result'.[117] What is risky about the certainty of the event occurring or the unpredictability of it is '[o]nly the length of time that will elapse'.[118] In the context of the *Danube Dam Case*, the ICJ argued, for instance, that 'dangers ascribed to the upstream reservoir were mostly of a long-term nature and, above all, that they remained uncertain'.[119] That is, the Court was more concerned in this case with establishing the certainty of harm than worrying about the risk of whether it would occur or not sometime in the future.

Clearly, the Court was not dismissing the need to prevent environmental harm. At one point it said that:

> [i]n the field of environmental protection, vigilance and prevention are required on account of the often irreversible character of damage to the environment and of the limitations inherent in the very mechanism of reparation of this type of damage.[120]

---

[114] On this important distinction between certain and uncertain risk see Marjolein van Asselt and Ellen Vos, 'The Precautionary Principle and the Uncertainty Paradox' (2006) 9(4) *Journal of Risk Research* 313. On the difference it makes in the context of legal disputes see Nicolas de Sadeleer, *Environmental Principles: From Political Slogans to Legal Rules* (2005); Nicolas de Sadeleer, 'The Effect of Uncertainty on the Threshold Levels to which the Precautionary Principle Appears to be Subject' in Maurice Sheridan and Luc Lavrysen (eds), *Environmental Law Principles in Practice* (2002) 32. See also Helga Nowotny, Peter Scott, and Michael Gibbons, *Re-Thinking Science: Knowledge and the Public in an Age of Uncertainty* (2001).

[115] van Asselt and Vos, above n 114, 314; de Sadeleer, *Environmental Principles*, above n 114, 156–161.

[116] *Memorial of the Republic of Hungary*, above n 79, 201–203.

[117] de Sadeleer, *Environmental Principles*, above n 114, 158.

[118] Ibid, 159. See also de Sadeleer, 'The Effect of Uncertainty on the Threshold Levels to which the Precautionary Principle Appears to be Subject', above n 114, 32–36.

[119] *Case Concerning the Gabčíkovo-Nagymaros Project (Hungary/Slovakia)* [1997] ICJ Rep 7, [55].

[120] Ibid, [140].

Nonetheless, it did adopt an interpretation of 'imminent' and 'peril' which favoured prevention as opposed to precaution. It seemed to have ignored the debates and discussions relating to 'uncertainty'.[121] This suggests that the precautionary principle has yet to have a productive effect on framing the power that the Court will exercise in determining issues before it.

It appears from this discussion that the Court was being conservative about its own role in the resolution of disputes involving claims of uncertain risk to the environment. Its interpretation of Article 33(1)(a) of the *Draft Articles on State Responsibility* as requiring certain risk in relation to establishing a 'peril' meant that the Court was also restricting any opportunities states might have to abuse the potential of this provision, which it claimed is now part of customary international law. It presumes the capacity of a state to be able to prove with certainty whether harm will occur or not in making out a case for ecological necessity. It is also arguable that the Court has changed the potential nature of the issues that future litigants may raise before it where they involve claims based on the precautionary principle. In other words, the Court through its interaction with the parties in this case was able to identify the potential challenge of leaving itself vulnerable to determinations of scientific uncertainty where it might deliver inconsistent decisions between cases before it.

6.5.2 Equitable Utilisation and Sustainable Development

In 1997 Hungary, in its submission to the ICJ in the *Danube Dam Case*, argued that Variant C took away between 40 and 43 per cent of the 50 per cent of the waters of the Danube that would otherwise be available for its use.[122] Slovakia had argued however that Variant C was a 'justified countermeasure to Hungary's illegal acts'.[123] The ICJ responded to Slovakia's argument by declaring that it had deprived Hungary of its right equitably to use the Danube as a natural resource. The Court also noted that one of the legal effects of its decision was that in interpreting Articles 15, 19 and 20 of the 1977 Treaty it had to 'look afresh at the effects on the

---

[121] For a good summary discussion of uncertainty and risk in relation to the use of the precautionary principle in international law see Arie Trouwborst, *Precautionary Rights and Duties of States* (2006).

[122] *Memorial of the Republic of Hungary*, above n 79, 229. The Court itself notes that 'Variant C led Czechoslovakia to appropriate, essentially for its use and benefit, between 80 and 90 per cent of the waters of the Danube before returning them to the main bed of the river': *Case Concerning the Gabčíkovo-Nagymaros Project* [1997] ICJ Rep 7, [78].

[123] Ibid, [85].

environment of the operation of the Gabčíkovo power plant' using 'new norms' and 'new standards'.[124] This discussion looks more closely at the Court's declaration relating to the lack of proportionality in Slovakia's countermeasures against Hungary.

Hungary had argued that the diversion had deprived it of its right to 50 per cent of the flow of the Danube. That is, by polluting the watercourse, Slovakia had compounded the degree to which it lost its equitable share of the resource.[125] Slovakia had argued however that Variant C was in fact the optimal way to use the Danube, given that the relevant circumstances were brought about by Hungary's intentions to terminate the Treaty.[126] Importantly, Slovakia drew on Article 6 of the 1991 ILC draft articles on the law of non-navigational uses of international watercourses[127] to argue that the relevant criteria for determining what is equitable and reasonable included ones relating to the social and economic impacts of the diversion rather than the environmental ones.[128] This was in total contrast

---

124 Ibid, [140].

125 *Memorial of the Republic of Hungary*, above n 79, 229–230.

126 *Memorial of the Slovak Republic in the Case Concerning the Gabčíkovo-Nagymaros Project (Hungary v Slovakia)* Vol 1 (2 May 1994), [302]–[304], www.icj-cij.org/docket/index.php?p1=3&p2=3&k=8d&case=92&code=hs&p3=1.

127 *Draft Articles on the Law of the Non-Navigational Uses of International Watercourses* (1991) 2(2) *Yearbook of the International Law Commission* 66 ('1991 ILC *draft articles on watercourses*'). These draft articles were amended in 1994 and again in 1997 before the UNGA approved them by 103 votes (with 3 against and 27 abstentions) as the *United Nations Convention on the Law of Non-Navigational Uses of International Watercourses*, opened for signature 21 May 1997, (1997) 36 ILM 700 (not in force) ('*Watercourses Convention*'). On the history and analysis of the agreement see Stephen McCaffrey and Robert Rosenstock, 'The International Law Commission's Draft Articles on International Watercourses: An Overview and Commentary' (1996) 5(2) *Review of European Community and International Environmental Law* 89; Stephen McCaffrey, *The Law of International Watercourses* (2nd ed, 2007) 217. On international watercourses more generally see Patricia Wouters (ed), *International Water Law: Selected Writings of Professor Charles B Bourne* (1997); Eyal Benvenisti, 'Collective Action in the Utilization of Shared Freshwater: The Challenges of International Water Resources Law' (1996) 90 *The American Journal of International Law* 384; Stephen McCaffrey and Mpazi Sinjela, 'The 1997 United Nations Convention on International Watercourses' (1998) 92 *American Journal of International Law* 97.

128 Its submission to the Court at para 7.8 states that: '[w]hether on the basis of (b) "the social and economic needs of the watercourse state concerned", (c) "existing and potential uses of the watercourse", or (f) "the availability of alternatives, or corresponding value, to a planned or existing use", Variant "C" is clearly an equitable and reasonable utilisation, and lawful.': *Memorial of the Slovak Republic*, above n 126, 304.

to Hungary's arguments, which relied on environmental considerations relating to the equitable and reasonable use of the Danube as a resource.

The portion of the Court's judgment in the *Danube Dam Case* which is discussed in this section is extracted here to facilitate a closer reading of it. The Court wrote:

> In the view of the Court, an important consideration is that the effects of a countermeasure must be commensurate with the injury suffered, taking account of the rights in question.
>
> In 1929, the Permanent Court of International Justice, with regard to navigation on the River Oder, stated as follows:
>
> > '[the] community of interest in a navigable river becomes the basis of a common legal right, the essential features of which are the perfect equality of all riparian States in the user of the whole course of the river and the exclusion of any preferential privilege of any one riparian State in relation to the others' . . . .
>
> Modern development of international law has strengthened this principle for non-navigational uses of international watercourses as well, as evidenced by the adoption of the Convention of 21 May 1997 on the Law of the Non-Navigational Uses of International Watercourses by the United Nations General Assembly.
>
> The Court considers that Czechoslovakia, by unilaterally assuming control of a shared resource, and thereby depriving Hungary of its right to an equitable and reasonable share of the natural resources of the Danube – with the continuing effects of the diversion of these waters on the ecology of the riparian area of the Szigetkoz – failed to respect the proportionality which is required by international law.[129]

The Court in this part of its judgment endorses the idea, initially developed in the *Case Relating to the Territorial Jurisdiction of the International Commission of the River Oder*,[130] that there is a 'common legal right' to a 'community of interest' in a 'navigable river'.[131] It claimed that this right had been strengthened through '[m]odern' developments of it 'in international law'.[132] Slovakia could not have 'unilaterally' assumed 'control of a shared resource', namely the Danube, and as a result deprived 'Hungary of its right to an equitable and reasonable share of the natural resources

---

129 *Case Concerning the Gabčíkovo-Nagymaros Project (Hungary/Slovakia)* [1997] ICJ Rep 7, [85].

130 [1929] PCIJ (ser A) No 23, [27].

131 *Case Concerning the Gabčíkovo-Nagymaros Project (Hungary/Slovakia)* [1997] ICJ Rep 7, [85]. On the concept of a 'community of interest' and its modern relevance see McCaffrey, above n 127, 147–167.

132 *Case Concerning the Gabčíkovo-Nagymaros Project (Hungary/Slovakia)* [1997] ICJ Rep 7, [85].

of the Danube'.[133] The Court had earlier in its judgment also confirmed the importance and relevance of the idea of equitable and reasonable use when it declared that Hungary had not 'forfeited its basic right to an equitable and reasonable sharing of the resources of an international watercourse' just because it had violated its terms by not continuing to implement them.[134] In this sense, the Court simply confirmed the presence of a right to an equitable and reasonable share of a natural resource and acknowledged the rights states had not to have it unilaterally taken away from them by a state upstream to them.

In highlighting the presence of this right to equitably and reasonably use an international watercourse, the Court referred to the 1997 *Watercourses Convention*.[135] However it did not make explicit reference to Article 5 which actually codifies the idea that they draw upon.[136] In relation to the Court's pronouncement of the disproportionate nature of the counter-measure taken by Slovakia, Higgins has written that the:

> Court must be taken to have linked the unilateral control to its finding that Hungary was deprived of an equitable and reasonable share of the natural resource. And it appears to have determined that such a deprivation cannot be proportionate to any prior illegality relating to the watercourse (and not just that the deprivation caused in the instant case was disproportionate to the prior illegalities of suspension and abandonment).[137]

What is interesting about this observation is that there is nothing wrong with a state on its own deciding how to use an international river as long as

---

133 Ibid

134 Ibid, 78.

135 Opened for signature 21 May 1997, (1997) 36 ILM 700 (not in force). In relation to the significance of this convention and an analysis see in particular, Part IV of the work of Stephen McCaffrey, above n 127. This agreement helps to define the basic rights and duties relating to the non-navigational uses of freshwater resources shared between states. In situations where states have already agreed on how they will share freshwater resources, the 1997 *Watercourses Convention* usually provides them with provisions to help interpret their bilateral agreements.

136 *Case Concerning the Gabčíkovo-Nagymaros Project (Hungary/Slovakia)* [1997] ICJ Rep 7, [85]. It is interesting to note that Slovakia in its argument made extensive reference to the 1991 ILC *draft articles on watercourses,* above n 127, instead of the much later ones: *Memorial of the Slovak Republic*, above n 126, 302. The ICJ did in fact refer to Art 5(2) of the 1977 *Watercourses Convention* in a later part of its judgment: *Case Concerning the Gabčíkovo-Nagymaros Project (Hungary/Slovakia)* [1997] ICJ Rep 7, [147].

137 Rosalyn Higgins, 'Natural Resources in the Case Law of the International Court' in Alan Boyle and David Freestone (eds), *International Law and Sustainable Development: Past Achievements and Future Challenges* (1999) 101, 110.

its decision does not impact on the potential of other states to the equitable use of the same international watercourse.[138] What is therefore at issue here, which Higgins does not really get to, is what was Hungary deprived of that was so significant to have made the unilateral decision of Slovakia so meaningful for the Court. It is therefore clear that the group of judges deciding the case must have had something else at the back of their minds when making this rather difficult and yet crucial pronouncement in terms of the rights and interests of Slovakia in particular. That is, it could not have been the concern that Slovakia had decided on its own to take away Hungary's right to a natural resource. McCaffrey also confirms the vagueness of the ICJ's approach when he voiced his interest in the way the Court rejected Slovakia's claim to having taken a proportionate countermeasure. He has written that 'it is remarkable that the Court seemed to believe it was obvious that Slovakia's diversion was per se a disproportionate response to Hungary's internationally wrongful act'.[139] McCaffrey is suggesting that the Court must have weighed Slovakia's actions as a contrast to their disadvantages for Hungary.

This is important because it has been argued that either actual or potential trans-frontier harm is not in itself problematic unless a state can show that because of this it has lost its right to an equitable and reasonable utilisation of the international watercourse.[140] The Court certainly voiced its concern over the impact of Variant C on Hungary. It pointed out that the diversion of the Danube had 'continuing effects' on the ecology of the riparian area of the Szigetkoz'.[141] Nevertheless the Court avoided evoking the idea that 'no harm' must come to any watercourse state, and thereby gave priority to the idea of equitable and reasonable utilisation.[142] This is confirmed by the fact that the ICJ did not refer to Article 7 of the 1997 *Watercourses Convention* which is also important in the context of this discussion. Article 7 provides that 'watercourse States shall . . . . take all appropriate measures to prevent the causing of significant harm to other watercourse states'.

---

138 See, eg, Ibrahim Kaya, *Equitable Utilization: The Law of the Non-Navigational Uses of International Watercourses* (2003) 151–179.

139 McCaffrey, above n 127, 217.

140 Kaya, above n 138, 169.

141 *Case Concerning the Gabčíkovo-Nagymaros Project (Hungary/Slovakia)* [1997] ICJ Rep 7, [85].

142 For a discussion of 'equitable and reasonable utilisation' in the context of the 1997 *Watercourses Convention* see Charles B Bourne, 'The Primacy of the Principle of Equitable Utilization in the 1997 Watercourses Convention' (1997) 35 *Canadian Yearbook of International Law* 221.

To understand this rather puzzling but yet crucial part of the judgment of the Court[143] it is important to look at Article 5(1) of the 1997 *Watercourses Convention*. Article 5(1) provides that:

> Watercourse States shall in their respective territories utilize an international watercourse in an equitable and reasonable manner. In particular, an international watercourse shall be used and developed by watercourse States with a view to attaining optimal and *sustainable utilization* thereof and benefits therefrom, taking into account the interests of the watercourse States concerned, consistent with adequate protection of the watercourse.[144]

What is important about Article 5(1) is that in its draft form, namely the 1991 and 1994 ILC *draft articles on watercourses*, there was no reference to the term 'sustainable' in the context of what equitable and reasonable utilisation might mean to states. The preamble to the 1997 *Watercourses Convention* also identifies sustainable utilisation as an important goal to promote for the 'present and future generations' which was not present in the 1991 and 1994 version of the same instrument. It must be noted however that the 1997 *Watercourses Convention* had been adopted on 21 May 1997 by the General Assembly of the United Nations (UNGA) (four months before the case was heard by the Court), meaning that its status as customary law was still debatable and the instrument had certainly not come into force as a treaty at the time the ICJ was referring to it. Despite these comments, the Convention did reflect around 20 years of work by the ILC.[145]

What stands out in Aritcle 5(1) is the idea that in order equitably and reasonably to use an international watercourse a state must take the sustainable and optimal use of the natural resource into account. Article 5(1) does not say that this has to be an explicit part of the consideration, but a meaningful assessment of equitable use cannot ignore sustainability. Some have argued that the concept of equitable and reasonable use of a watercourse is synonymous with the idea of using it sustainably.[146] Importantly, considerations of sustainability potentially add new elements to what is equitable and reasonable in the sense that it emphasises the needs of the future generation. That is, one big difference between equitable and sustainable utilisation is the presence of the future generation in structuring

---

143 The Court's finding that Slovakia had acted disproportionately meant that it failed in its main claim for damages against Hungary.

144 Emphasis added.

145 McCaffrey, above n 127, 216.

146 See, for instance, Kaya, above n 138, 179–183.

the Court's power over the litigation. In fact before the 1997 *Watercourses Convention* was approved by the UNGA, writers had argued that the concept of equitable utilisation and the criteria for determining it did not value the future generation properly.[147] The reference to the future generation in the preamble to the 1997 *Watercourses Convention* also confirms this new priority of the agreement as highlighted in Article 5(1).

From this one might conclude that the unsustainability of Variant C is in the impact it would have had on the future generation because of the effects not only of the water quality of the river but the riparian areas of Szigetkoz, which the Court itself highlighted in its judgment.[148] As a result, the unilateral control of the river is significant in terms of its implications for the future generation. This conclusion, although not obvious on an immediate reading of paragraph [85] of the judgment, is confirmed, implicitly, in paragraph [140] when the Court wrote that 'this need to reconcile economic development with protection of the environment is aptly expressed in the concept of sustainable development'. More importantly it continued to say that:

> For the purposes of the present case, this means that the Parties together should look afresh at the effects on the environment of the operation of the Gabčíkovo power plant. In particular they must find a satisfactory solution for the volume of water to be released into the old bed of the Danube and into the side-arms on both sides of the river.[149]

This reference to needing to reconsider their operations in the light of concerns about sustainable development adds support to the observations above about why the Court might have decided that Slovakia's diversion of the Danube was disproportionate as a countermeasure to what Hungary had done. The inability to properly formulate what the rights of the future generation will be worth might have given enough reason to the Court for avoiding a discussion of it in its judgment.

Sustainable development in its interplay with the power of the Court produced an alternative possibility for the application of the equitable utilisation of international watercourses in the *Danube Dam Case*. It is accurate to suggest that the Court learnt through its interaction with Hungary and Slovakia that the articulation of sustainability in Article 5(1)

---

147 See Mete Erdem, 'Sustainable Utilisation of International Watercourses: a Legal Overview' (1992) 12 *Istanbul University International Law Bulletin* 41.

148 *Case Concerning the Gabčíkovo-Nagymaros Project (Hungary/Slovakia)* [1997] ICJ Rep 7, [85].

149 Ibid, [140].

of the 1997 *Watercourses Convention* was necessary. This is important, given the lack of apparent connections before this case between the concepts of equitable utilisation and sustainable development.[150] Sustainable development appears to have functioned not as a norm with a particular fixed meaning but as an episteme that gave context and direction to what the Court was being asked to do in relation to the equitable use of an international watercourse. The power of the Court was disciplined in favour of giving extra protection to the environment in circumstances that could have just as easily tipped in favour of the social and economic priorities of managing watercourses. In this sense, sustainable development appears to have heightened the awareness of the Court in favour of existing norms that states had adopted which favoured the environmental dimension of the use of international rivers.

In contrast to the above discussion in section 6.5.1 on the precautionary principle, it is arguable that sustainable development had more of a productive effect on the Court, because it interpreted equitable utilisation in favour of environmental concerns raised by Hungary. This is in contrast to its interpretation of 'imminent peril', which did not favour the more environmentally sensitive approach that could have been shaped by the Court had the precautionary principle had more of an effect on it. However, the differences between the two situations are important in terms of the way that the Court's power over Hungary and Slovakia was structured. In assessing what 'imminent peril' meant, the Court did not appear to have any other evidence from, for instance, the *Draft Articles on State Responsibility* which would have suggested that the practices of states favoured giving credence to 'uncertain' risks to the natural environment. In interpreting what 'equitable utilisation' of a watercourse referred to in assessing what Hungary had been deprived of, the Court instead had the 1997 *Watercourses Convention* which, although it had not come into force at the time the Court was delivering its judgment, was reflective of years of drafting work by the ILC. In other words, the approach of the Court to how it assessed whether Slovakia had deprived Hungary of its right to equitable utilisation of the Danube was socially constructed through the productive effect that sustainable development had on the Court's power. Article 5(1) of the 1997 *Watercourses Convention* clearly legitimated, for the Court, concerns about sustainability in relation to the use of an international watercourse like the Danube.

---

[150] Alan Boyle, 'The *Gabčíkovo-Nagymaros Case*: New Law in Old Bottles' (1997) 8 *Yearbook of International Environmental Law* 13.

6.6 PRODUCTIVE POWER OF PROTECTING THE FUTURE GENERATION AND TRANSBOUNDARY HARM

The discussion in this section continues to explore the interplay between principles and productive power, but from a different perspective. It examines whether the open-textured nature of some environmental principles can function as frames which help the ICJ to develop a common understanding of deeper philosophical views on how the environment should be protected. The significance of this discussion is to highlight how the ICJ can potentially develop unique positions on environmental issues which can then be used in litigation before it as well as in the development of international law more generally.

In the *Corfu Channel Case (United Kingdom v Albania)* the ICJ noted that 'certain general and well-recognized principles' were binding on Albania, one of which was that every state had an 'obligation not to allow knowingly its territory to be used for acts contrary to the rights of other States'.[151] Although the *Corfu Channel Case* did not involve an environmental dispute, various scholars use the transboundary harm principle identified by the Court to indicate that as far back as 1949 the ICJ had made statements 'pertinent to transboundary environmental issues'.[152] The transboundary harm principle in its different formulations has been described as the 'cornerstone of international environmental law'.[153]

---

151 [1949] ICJ Rep 4 ('*Corfu Channel Case*'). This norm will hereinafter be referred to as the 'transboundary harm principle'.

152 Sands, above n 10, 629; Nathalie Horbach and Pieter Bekker, 'State Responsibility for Injurious Transboundary Activity in Retrospect' (2003) 50 *Netherlands International Law Review* 327, 343–344. On this point see also the dissenting opinion of Judge Weeramantry V-P in *Request for an Examination of the Situation in Accordance with Paragraph 63 of the Court's Judgment of 20 December 1974 in the Nuclear Tests (New Zealand v France) Case* [1995] ICJ Rep 288 where at 362 he noted that the:

> *Corfu Channel case* laid down the environmentally important principle that, if a nation knows that harmful effects may occur to other nations from facts within its knowledge and fails to disclose them, it will be liability to the nation that suffers damage.

153 Philippe Sands, *Principles of International Environmental Law* (2nd ed, 2003) 236; Xue Hanqin, *Transboundary Damage in International Law* (2003); Günther Handl, 'Transboundary Impacts' in Daniel Bodansky, Jutta Brunnée and Ellen Hey (eds), *The Oxford Handbook of International Environmental Law* (2007) 531. However, others point to the fact that the principle is simply about protecting the territorial integrity of a sovereign: Klaus Bosselmann, 'Environmental

It was originally raised as a norm in the *Trail Smelter Case (Canada/United States of America)* arbitration decision involving Canada and the United States.[154] Although not in the form as stated by the ICJ in later cases, it was codified in Principle 21 of the *Stockholm Declaration* and then in similar terms by Principle 2 of the *Rio Declaration*.[155] The version of the Principle as it appears in Principle 21 of the *Stockholm Declaration* identifies that:

> States have, in accordance with the Charter of the United Nations and the principles of international law, the sovereign right to exploit their own resources pursuant to their own environmental policies, and the responsibility to ensure that activities within their jurisdiction or control do not cause damage to the environment of other States or of areas beyond the limits of national jurisdiction.[156]

It has been argued that if the *Trail Smelter Arbitration* were to be retried it would not be decided any differently given the way international environmental law has evolved since that case.[157] Interestingly, in the *Trail Smelter Arbitration*, Canada did not have to stop operating the smelter that caused the pollution, but rather had to minimise any harm that it caused to the territory of the United States. Horbach and Bekker argue that because of the precautionary principle the onus of proof on states in the light of the transboundary harm principle may in fact change.[158] As a result some types of domestic activities could now cause 'certain types of significant transboundary harm (biosafety or biodiversity injury)',

---

Governance: A New Approach to Territorial Sovereignty' in Robert Goldstein (ed), *Environmental Ethics and Law* (2004) 293, 301.

[154] *Trail Smelter Case (Canada/United States of America)* (1938 and 1941) 3 RIAA 1911. On this topic see, more generally, Rebecca Bratspies and Russell Miller (eds), *Transboundary Harm in International Law: Lessons from the Trail Smelter Arbitration* (2006).

[155] For discussions of the differences between Principle 21 of the *Stockholm Declaration* and Principle 2 of the *Rio Declaration* and the IJC's formulation see also Brown Weiss, above n 18.

[156] Principle 2 of the *Rio Declaration* encourages states to exploit 'their own resources pursuant to their own environmental and developmental policies'. The inclusion of the term 'developmental policies' in Principle 2 has been criticised; see, for instance, David Wirth, 'The Rio Declaration on Environment and Development: Two Steps Forward and One Back, or Vice Versa' (1995) 29 *Georgia Law Review* 599, and Marc Pallemaerts, 'International Environmental Law from Stockholm to Rio: Back to the Future' (1992) 1 *Review of European Community and International Environmental Law* 254.

[157] Horbach and Bekker, above n 152, 367–370.

[158] Ibid, 370.

whether it is today or some time in the future, which may need to be stopped. [159] They also point out that this view is consistent with a reading of Article 1 of the 2001 *Draft Articles on the Prevention of Transboundary Harm from Hazardous Activities* adopted by the ILC.[160] This observation by Horbach and Bekker suggests that the transboundary harm principle appears to have a core which has not been modified but that its application has been further nuanced to account for an increasing awareness of the risk associated with certain types of activities that are not unlawful under international law.[161]

The ICJ has confirmed a version of the transboundary harm principle which not only reproduces a core idea common amongst many definitions of the norm, but because of its open-textured nature has produced it in a new form by drawing on a variety of novel ideas and concepts. Given that this formulation is similar to but also different from what appears for instance in the 2001 ILC *Draft Articles on Transboundary Harm*, or earlier in Principle 21 of *Stockholm Declaration* or Principle 2 of the *Rio Declaration*, it is arguable that the social learning is uniquely that of the ICJ. It is however the interaction of the ICJ with the productive power of the idea of needing to protect the future generation that has provided further nuance to the open-textured nature of the transboundary harm principle in this instance.[162] As a result the open-textured nature of the

---

159 Ibid.

160 *Report of the International Law Commission, 53rd Session*, UN Doc A/56/10 (2001) ('ILC *Draft Articles on Transboundary Harm*'). Art 1 outlines the scope of the Convention, stating that '[t]he present draft articles apply to activities not prohibited by international law which involve a risk of causing significant transboundary harm through their physical consequences'. See also Prue Taylor, *An Ecological Approach to International Law: Responding to challenges of climate change* (1998) especially ch 4.

161 This is also consistent with the historical reasons why the ILC separated its discussion of the rule of state responsibility which it later codified in the *Draft Articles on Responsibility of States for Internationally Wrongful Acts, Report of the International Law Commission, 53rd Session*, UN Doc A/56/10 (2001) (noted in GA Res 56/83, UN Doc A/RES/56/83 (2001)), and those acts that are not unlawful but which have injurious consequences for states: *Draft Articles on the Prevention of Transboundary Harm from Hazardous Activities, Report of the International Law Commission, 53rd Session*, UN Doc A/56/10 (2001).

162 As discussed in chapter 2, this book uses the notion of protecting the future generation as an idea that is collectively held amongst groups of actors. It is its instantiation in norms like the transboundary harm principle which confirms the idea that it is part of the collective identity of actors more generally: see Joseph DesJardins, *Environmental Ethics: An Introduction to Environmental Philosophy* (4th ed, 2005) especially ch 4. See also Sagoff, above n 50. For an alternative view in international relations see Stone, above n 50.

transboundary harm principle has allowed the ICJ to establish how it identifies with the idea of protecting the future generation from environmental harm. The rest of this section will show how the version of the transboundary harm principle developed and now adopted by the ICJ through a number of cases is unique, and how its construction is a reflection of the interplay between the productive power deployed within the ICJ and the principle.

In 1994 the UNGA followed the lead of the World Health Organisation (WHO) a year before it had asked the ICJ for an Advisory Opinion on the following question: '[i]s the threat or use of nuclear weapons in any circumstance permitted under international law?'.[163] Amongst other matters, the Court had therefore to consider whether the various formulations of the transboundary harm principle that had been put to it would be violated were nuclear weapons to be used by any state.[164] In delivering its judgment in the *Nuclear Weapons Case* the ICJ also developed its own language for giving expression to the transboundary harm principle when it stated that:

> the environment is not an abstraction but represents the living space, the quality of life and the very health of human beings, including generations unborn. The existence of the general obligation of States to ensure that activities within their jurisdiction and control respect the environment of other States or of areas beyond national control is now part of the corpus of international law relating to the environment.[165]

The ICJ has since confirmed this exact expression of the principle in important cases such as the *Danube Dam Case*[166] and most recently when it provided *Provisional Orders* in the *Case Concerning Pulp Mills in the River Uruguay (Argentina v Uruguay)*.[167] Arguably, it is now in a form which, because of its repetition in several judgments, can be used by others in the future when they litigate before the Court. Although it is possible

---

163 *Legality of the Threat or Use of Nuclear Weapons (Advisory Opinion)* [1996] ICJ Rep 226, [1].

164 Ibid, [27].

165 Ibid, [29]. On this case more generally see John Burroughs, *The (Il)legality of Threat or Use of Nuclear Weapons: A Guide to the Historic Opinion of the International Court of Justice* (1997).

166 *Case Concerning the Gabčíkovo-Nagymaros Project (Hungary/Slovakia)* [1997] ICJ Rep 7, [140].

167 *Case Concerning Pulp Mills in the River Uruguay (Argentina v. Uruguay), Request for the Indication of Provisional Measures,* Order of 13 July 2006, 45 ILM 1025, [72] ('*River Uraguay Case (Provisional)*').

to suggest that the version of the transboundary harm principle is unique to the facts of the *Nuclear Weapons Case*,[168] its repetition using the exact formulation in subsequent judgments suggests that this is now a debatable issue.

A number of differences can be observed between the two different versions of the transboundary harm principle developed by the ICJ in the *Corfu Channel Case* and the *Nuclear Weapons Advisory Opinion*. These differences include:

1. the focus of the principle is now not just on sovereign territorial space but also areas of the world that are beyond national control;[169]
2. the idea that states must 'respect' as well as not harm the environment. This formulation requires a level of care from states beyond simply preventing harm;[170] and
3. an explicit statement that the environment includes the interests that the future generation might have in the 'living space' of humans and that we must 'ensure that they have some quality to their lives and can remain healthy'.

In relation to the first point, the original formulation of the transboundary harm principle did not include references to the global commons in any way.[171] Although Principle 21 of the *Stockholm Declaration* and Principle 2 of the *Rio Declaration* refer to areas that are beyond the 'limits of national jurisdiction' the ICJ specifies more precisely that the parts of the world that are beyond territorial boundaries are important for the purposes of the transboundary harm principle. The ICJ's interpretation of this idea, however, is limited to areas that are a 'living space' for humans or which could impact on our health in some way. Importantly, this formulation by the ICJ appears to go further than what is specified in the 2001 ILC *Draft Articles on Transboundary Harm*.[172] In Article 2(c) for instance it limits the

---

[168] See, eg, Stephens, above n 10, 143, fn 150, has suggested that Principle 21 of the *Stockholm Declaration* could not have been used because nuclear weapons could cause much harm to the natural environment of the world and not just to those of neighbouring states.

[169] See, eg, Kathy Leigh, 'Liability for Damage to the Global Commons' (1992) 14 *Australian Year Book of International Law* 129.

[170] For another examination of the use of the word 'respect' by the ICJ see Brown Weiss, above n 18.

[171] See, for instance, Leigh, above n 169.

[172] *Draft Articles on the Prevention of Transboundary Harm from Hazardous Activities, Report of the International Law Commission, 53rd Session*, UN Doc A/56/10 (2001).

scope of its provisions to harm caused by states to places that are 'under the jurisdiction or control of a State'.[173] This extension of the principle by the ICJ stands out, given that it has been commented that the ILC was 'cautious to develop' the articles in the Convention 'along the lines established by related international instruments, in particular the principles embodied in the Rio Declaration of 1992, the Stockholm Declaration of 1972'.[174] The implication of the extended transboundary harm principle in this way by the ICJ is important because it can potentially apply to situations that involve multiple actors and include harm to the global commons, such as the depletion of the ozone layer.[175]

Secondly, the ICJ raises the prospect that the word 'respect' in the context of the transboundary harm principle changes the threshold required for assessing what states are allowed to do to areas outside their jurisdiction.[176] The commentary on the 2001 ILC *Draft Articles on Transboundary Harm* identifies the terms 'significant', 'serious', and 'substantial' as some of the words used in judgments and treaties to describe existing thresholds for assessing whether the damage caused to territory outside a state's jurisdiction should attract liability.[177] The 2001 ILC *Draft Articles on Transboundary Harm* actually adopted the term 'significant transboundary harm' as the threshold for what states should avoid.[178] In its commentaries to the Convention the ILC noted that the term 'significant' involves

---

[173] Art 2(c) provides that '[t]ransboundary harm' means harm caused in the territory of or in other places under the jurisdiction or control of a State other than the State or origin, whether or not the States concerned share a common border'.

[174] Bruno Simma, 'The Work of the International Law Commission at Its Fifty-Third Session (2001)' (2002) 71 *Nordic Journal of International Law* 123, 138.

[175] See Pemmaraju Sreenivasa Rao, 'Environment as a Common Heritage of Mankind: a Policy Perspective' in United Nations, *International Law on the Eve of the Twenty-First Century: Views from the International Law Commission: réflexions de codificateurs* (1997) 201, 213, referred to in Horbach and Bekker, above n 152, fn 49. See also Pemmaraju Sreenivasa Rao, 'International Liability for Transboundary Harm' (2004) 34(6) *Environmental Policy and Law* 224, 227.

[176] The ICJ in the *Nuclear Weapons Case* at [30] also used the word 'respect' in the context of the necessity and proportionality of measures aimed at self-defence. It said that '[r]espect for the environment is one of the elements that go to assessing whether an action is in conformity with the principles of necessity and proportionality'.

[177] International Law Commission, 'Commentary on the Draft Articles on Prevention of Transboundary Harm from Hazardous Activities' (2001) 2(2) *Yearbook of the International Law Commission* 144, 152.

[178] Art 3 of the *Draft Articles on the Prevention of Transboundary Harm from Hazardous Activities, Report of the International Law Commission, 53rd Session*, UN Doc A/56/10 (2001) requires that '[t]he State of origin shall take all

'more factual considerations than legal determination'.[179] It noted that the term 'significant', while determined by factual and objective criteria, also involved 'a value determination which depends on the circumstances of a particular case and the period in which such determination' was to be made.[180] Clearly, the term 'respect' is a much more open-textured notion than the words 'do not cause' which appear in Principle 21 of the *Stockholm Declaration* and is far more value driven than 'significant' which is used in Article 3 of the 2001 ILC *Draft Articles on Transboundary Harm*.[181] The reason for making this distinction is that the ICJ leaves open the possibility that in arguing before it a state might be able to include for instance, a broader 'category of environmental expenses, including monitoring and assessment expenses and actual clean-up costs'.[182] Brown Weiss has also argued that the use of the word 'respect' rather than 'do not cause' imposes a broader obligation on states than is required by Principle 21 of the *Stockholm Declaration* and Principle 2 of the *Rio Declaration*.[183]

Lastly, from a philosophical or ethical perspective the ICJ appears to be stating that our relationship with the environment should continue to be unashamedly anthropocentric.[184] The basis for acting to prevent transboundary harm is simply the particular value that nature has for human beings. This view of the principle has an impact on the direct or indirect responsibility states have to protect the natural world within another jurisdiction or globally if it has apparent value for human beings. Although this sounds critical, this approach of the ICJ to the transboundary harm principle is important for highlighting the way that its power has been produced by its concern over the future generation. That is, the rights of the future generation are in themselves an abstraction which needs to be

---

appropriate measures to prevent significant transboundary harm or at any event to minimize the risk thereof'.

[179] International Law Commission, above n 177, 152.

[180] Ibid, 153.

[181] The *Trail Smelter Case (Canada/United States of America)* (1938 and 1941) 3 RIAA 1911 used the words 'serious consequence'.

[182] Horbach and Bekker, above n 152, 367, use these words to describe the significance of the work of Panel 'F4' of the Panels of Commissioners for the United Nations Compensation Commission (www.unog.ch/uncc) in relation to the environmental issues before them. On the United Nations Compensation Commission more generally see Richard Lillich (ed), *The United Nations Compensation Commission* (1995).

[183] See Brown Weiss, above n 18, 340.

[184] Anthropocentrism values the environment in terms of what it offers to human beings. On the ethical approaches to international environmental law see Alexander Gillespie, *International Environmental Law, Policy and Ethics* (2000).

specified. In valuing nature in terms of a 'living space' as well as for the 'quality of life and the very health of human beings' the ICJ appears to be specifying the areas where emphasis must be placed in terms of the transboundary harm principle. Importantly, the ICJ also suggests that states are no longer just concerned with the impact that transboundary harm has on the sovereignty of other states, but the people and their relationship to the natural environment.

The above discussion shows that the ICJ commonly identifies with a particular approach to the transboundary harm principle in international law and politics which it has reaffirmed over a few of its judgments, including the *Nuclear Weapons Case*, the *Danube Dam Case* and also the *River Uruguay Case (Interim)*. There are significant differences between its approach to the principle and, for instance, the codification of it in the 2001 ILC *Draft Articles on Transboundary Harm*. Its power over the normative solutions to what in chapter 3 was referred to as the *protasis* or the 'scope of the rule' has been produced by other ideas and concepts such as the needs and interests of the future generation. The environmental principle in this case gave creative support to what the ICJ did in developing a common understanding within the Court of what transboundary harm means. In this process the ICJ has shown leadership internationally in defining more broadly some of the elements that states could rely upon in generating ideas about their rights and responsibilities. This is important because of the limited reach of the 2001 ILC *Draft Articles on Transboundary Harm*.

Functionally, the ICJ's particular conception of the transboundary harm principle has opened up the potential it has to encourage communication amongst actors before damage occurs. This is because of its reliance on the idea that states have to 'respect' the environment of others. That is, the value judgement involved in terms of what 'respect' means is significantly more subjective than the alternatives like 'significant' 'serious' or 'substantial'. A state is likely, under this conception of the principle, to have to engage in impact assessment of projects and conduct other due diligence activities in consultation with neighbouring states. This is especially needed if there is the potential that its activities within its jurisdiction might not show 'respect' towards those of the neighbouring state. In this sense, the ICJ was able to use the open-textured nature of the environmental principle to redefine it in, a way that still maintained its core connection with other formulations of it in, for instance, the 2001 ILC *Draft Articles on Transboundary Harm*. It was also able to provide a nuanced view of it that is more in line with protecting the interests of the people who live in neighbouring states whose environment might otherwise be harmed from a lack of due diligence by other states.

6.7 CONCLUSION

The ICJ has not frequently used environmental principles in determining the legal positions of parties appearing before it. This chapter has shown that this does not necessarily mean that they have not contributed to the way that the ICJ has positioned itself in changing international law and politics. In the three different ways that the role and function of environmental principles were examined it appears that the ICJ has been most directly engaged with the transboundary harm principle and has generated a common understanding of it that is significant in the context of international law more generally. Its open-textured nature meant that ideas about protecting the future generation were able to produce the Court's approach to it and distinguish its meaning from other efforts to regulate behaviour in relation to environmental harm that is not unlawful under international law. Although states have yet to use it directly in litigation before the Court it also has potential for creating expectations in international law and politics more generally. Also, given the potential discussed earlier in this work for other international courts and tribunals to draw from the jurisprudence of the ICJ, the collective social learning within the Court in this respect is significant for change amongst the jurisprudence of other courts and tribunals globally.

The analysis in this respect also highlighted how open-textured environmental principles which have a long history of developing in international law can create opportunities for the Court to shape and advance its own common understanding of environmental issues. Clearly, the fact that the ICJ has been willing to adopt a more environmentally focused view of the transboundary harm principle than the 2001 ILC *Draft Articles on Transboundary Harm* suggests that there is potential amongst the bench to learn socially through environmental principles apart from the broader codification of norms in conventions and treaties.

The approach of the Court however to whether environmental principles can structure how it approaches its interpretation of legal provisions and issues is not as promising. A closer examination of the pleadings of both Hungary and Slovakia in the *Danube Dam Case* showed that the parties commonly relied on environmental principles in making their arguments to the Court. The interactions amongst the parties in that case however failed to draw from the precautionary principle in shaping their views of 'imminent peril'. In contrast, in its approach to the concept of 'equitable utilisation', the Court appeared to have been influenced by the ideas which sustainability captures in its abbreviated form. This does not necessarily suggest that sustainable development played a significant role, given that Article 5(1) of the 1997 *Watercourses Convention* could have also shaped how the Court interpreted the concept of 'equitable utilisation'. However neither Hungary nor Slovakia had relied on the 1997 *Watercourses Convention* and did not argue in terms of sustainability.[185] It is therefore possible that concerns over sustainability had a diffuse effect on how the Court asserted itself in relation to Slovakia's diversion of the Danube as a countermeasure against Hungary.

The engagement of Hungary and Slovakia with the Court in the *Danube Dam Case* has created an expectation that it could once again emphasise environmental harm as an aspect of what is proportionality in the context of countermeasures. However, on a comparison of the interplay between the precautionary principle, sustainable development and productive power through the ICJ, it does not seem certain that environmental principles are in this respect likely to play a significant role in collective social learning within the bench. That is, the interpretations by the bench in the *Danube Dam Case* of established norms were not significantly swayed by either of these two principles. However, given the Court's approach to sustainability it is arguable that environmental principles may constitute how the Court interprets legal issues, especially if other treaties or conventions can confirm its approach. If international environmental law continues to develop in the light of concerns over the harm we cause to the natural environment, its principles could increasingly constitute how the Court exercises its interpretive powers over established norms. In this sense environmental principles could develop greater significance for learning within the Court and collective understanding emerging from it.

Lastly, it was argued that the social interaction of states post-adjudication is made meaningful by the Court's diffuse exercise of power, which may be ideologically framed, amongst other factors, by environmental principles. The process of litigating is in itself a social activity that helps to shape the preference of actors. Viewed in this way, environmental principles have the potential to encourage diplomatic relationships progressively to reveal what the actors' interests might be in the context of the overall dispute beyond the litigation itself. In these contexts environmental principles as abstract or open-textured norms are important because they function as norms with potential to communicate with parties beyond the case. In this sense environmental principles can identify for the parties a greater range of ideas and possibilities to use in their negotiations in contrast to rules that are narrowly constructed.

The work of the ICJ in dispute resolution is not an easy case in which

---

[185] Hungary's arguments were set against the damage that the diversion would cause to its environment and natural resources.

to study the role and function of environmental principles. Despite this the power of the Court appears to be constituted in different ways by the potential that environmental principles have to help it create meaning from its interactions with disputing litigants. In the context of the study in this chapter the abstract and open-textured nature of the environmental principles has meant that its function and role are subtler in their effects on the learning process.

7. The global compact, environmental principles and culture

7.1 INTRODUCTION

In December 2005 the 191 Member States of the United Nations General Assembly (UNGA) officially endorsed the Global Compact (GC) initiative which the former Secretary General, Kofi Annan, had established within his office in 2000.[1] In a much generalised but useful summary of its mission, the GC seeks to establish 'corporate citizenship amongst companies' in the world.[2] The GC is now a complex initiative which is coordinated by the Global Compact Office (GCO) and is part of the Secretary General's Office at the United Nations (UN).[3] As of September

---

1 *Towards global partnerships*, UN GAOR, 62nd Session, Agenda Item 61, A/C.2/62/L.33/Rev.1. See also Global Compact Office, *Letter to Global Compact Stakeholder* (2006) United Nations Global Compact, www.unglobalcompact.org/docs/how_to_participate_doc/Stakeholder_english_2008_final.pdf.

2 McKinsey & Company, *Assessing the Global Compact's Impact* (2004) United Nations Global Compact [1], www.unglobalcompact.org/docs/news_events/9.1_news_archives/2004_06_09/imp_ass.pdf. This study is a 'comprehensive impact assessment' of the GC that was commissioned by the Global Compact Office (GCO) in 2004 and completed on 11 May 2004. It does not actually define what it understands 'corporate citizenship' to mean in the context of the GC. On this point see also Surya Deva, 'Global Compact: A Critique of the UN's "Public-Private" Partnership for Promoting Corporate Citizenship' (2006) 34 *Syracuse Journal of International Law and Commerce* 107, 111–112. For the use of this terminology see also the report produced by the GCO and the Barcelona Centre for the Support of the Global Compact on the Local Networks: *Local Network Report* (2007) United Nations Global Compact 6, www.unglobalcompact.org/docs/news_events/8.1/LNReport_FINAL.pdf, which also refers to the GC as the 'largest voluntary corporate citizenship initiative' in the world. Others have summarised the purpose of the GC in different ways. For instance, King has suggested that 'in its simple form' it is about the 'dissemination of and adherence to good business practices': Betty King, 'The UN Global Compact: Responsibility for Human Rights, Labor Relations, and the Environment in Developing Nations' (2001) 34 *Cornell International Law Journal* 481, 482.

3 The review and actual changes to the GC framework are good illustrations of this complexity; see *Letter to Global Compact Stakeholder,* above n 1, 2.

2007 the GC had 4,600 participants and stakeholders from around 120 countries in the world.[4] This is compared to 2,900 participants and stakeholders in March 2006 when the GCO last reported on the growth of the initiative.[5] This is not to suggest that the GC is without its challenges or criticisms.[6] The 'symbolism of the Global Compact's creation and its established brand as a major initiative of the Secretary-General' is however 'surprisingly influential'.[7]

The core idea behind the GC initiative is to establish a set of ten principles which aim to influence the values of corporations in relation to human rights, labour, the environment and corruption and 'give a human face to the global market'.[8] The environmental principles that are a part of the GC are listed as the precautionary approach to environmental challenges; promoting greater environmental responsibility; and encouraging the development and diffusion of environmentally friendly technologies.[9] In

---

[4] Global Compact Office and the Barcelona Centre for the Support of the Global Compact, *Local Network Report*, above n 2, 6. See however David Weissbrodt, 'Business and Human Rights' (2005) 74 *University of Cincinnati Law Review* 55, who at 70 asks about 'the other 59,000 companies that are not covered by the Global Compact'.

[5] *Letter to Global Compact Stakeholder,* above n 1, 2.

[6] See, eg, Deva, above n 2; Maria Gjølberg and Audun Ruud, 'The UN Global Compact – A Contribution to Sustainable Development?' (Working Paper No 1/05, Centre for Development and Environment, University of Oslo, 2006). See also David Bigge, 'Bring on the Bluewash: a Social Constructivist Argument Against Using *Nike v Kasky* to Attack the UN Global Compact' (2004) 14 *International Legal Perspectives* 6, for a good summary of the major early criticisms of the GC, particularly in relation to companies and human rights.

[7] McKinsey & Company, above n 2. It is arguable that its success is partly due to many corporations using the GC as a branding exercise and for networking opportunities. For instance, the GCO in a report compiled in 2007 noted that 63 per cent of the 400 companies which they surveyed indicated that they had joined the GC to increase trust in their company. This is compared to 52 per cent that highlighted the fact that they wanted to address humanitarian concerns: Global Compact Office, *UN Global Compact Annual Review 2007 Leaders' Summit* (2007) United Nations Global Compact 11, www.unglobalcompact.org/docs/news_events/8.1/GCAnnualReview2007.pdf.

[8] United Nations, 'Secretary-General Proposes Global Compact on Human Rights, Labour, Environment, in address to World Economic Forum in Davos' (Press Release, 1 February 1999), www.un.org/News/Press/docs/1999/19990201.sgsm6881.html.

[9] See United Nations Global Compact, 'The Ten Principles' www.unglobalcompact.org/AboutTheGC/TheTenPrinciples/index.html. The other seven principles of the GC are: (1) 'Businesses should support and respect the protection of internationally proclaimed human rights'; (2) 'make sure that they are not complicit in human rights abuses'; (3) 'Businesses should uphold the freedom

2004 McKinsey & Company were commissioned by the GCO to complete an 'impact assessment' of the GC initiative.[10] This study singled out as the main focus for its empirical investigations the adoption by companies of the nine principles as they were in 2004.[11] It found that in the four years since its establishment, the:

> Global Compact has had noticeable, incremental impact on companies, the UN, governments and other civil society actors and has built a strong base for future results. The Compact has primarily accelerated policy change in companies, while catalysing a proliferation of 'partnership projects,' development-oriented activities that companies undertake with UN agencies and other partners. The Compact has also developed a solid participant base and local network structure, establishing itself as the largest voluntary corporate citizenship network of its kind.[12]

Figure 7.1, which is from the McKinsey & Company study, provides more perspective on these comments, in that 51 per cent of those surveyed said that the GC initiatives helped them to make the decision to engage with the principles easier as opposed to initiating their interest in them. Despite this, the study highlights that the principles used by the GC have the potential to socialise individual actors into changing their preferences in some way.

The difficulty of studying principles by simply asking whether companies or other stakeholders have complied with or internalised them is that one can easily ignore the abstract and open-textured nature of these norms.[13] The fact that one corporation has adopted a principle does not

---

of association and the effective recognition of the right to collective bargaining'; (4) 'the elimination of all forms of forced and compulsory labour'; (5) 'the effective abolition of child labour'; (6) 'the elimination of discrimination in respect of employment and occupation'; and (10) 'Businesses should work against all forms of corruption, including extortion and bribery': ibid.

10 McKinsey & Company, above n 2, 1.

11 A tenth principle dealing with anti-corruption was added to the other nine in 2004; see United Nations Global Compact, 'Transparency and Anti-corruption', www.unglobalcompact.org/abouttheGC/thetenprinciples/anti-corruption.html. As a secondary focus the study by McKinsey and Company, above n 2, 2 also examined the 'increased efficacy of the UN through a more collaborative approach to the private sector, support for governments seeking to spur a more effective role of business in society, and the convening of a unique multi-stakeholder network'.

12 Ibid, 2. It also noted criticisms, such as that the 'inconsistent participation and divergent and unmet expectations limit the impact on companies and continue to threaten the Compact's long-term credibility with participants'.

13 An alternative criticism of the McKinsey & Company study might relate to how they singled out the impact that the principles of the GC had on the

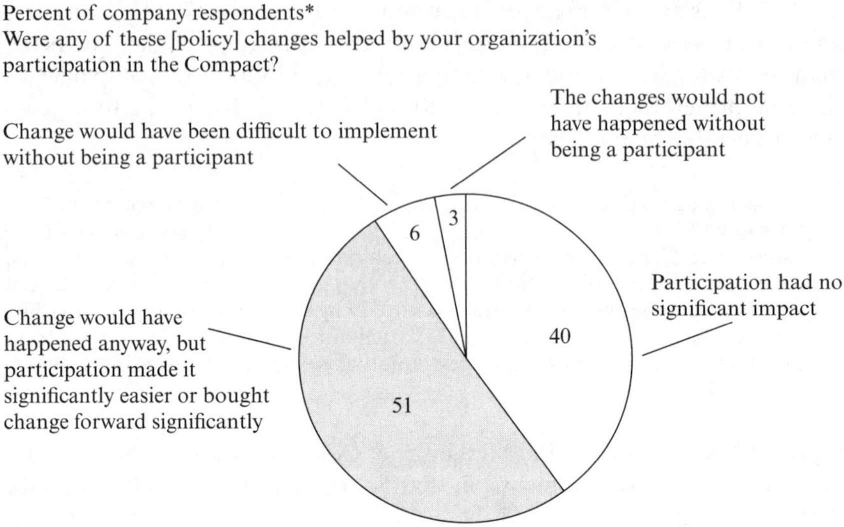

Notes: * Percent of respondents who indicated some change made since joining Compact (who made up 67% of total).
For most of the survey respondents who report that their companies made changes to implement the Compact's principles, participation in the Compact accelerated, rather than instigated the change.

Source: Mckinsey & Company, *Assessing the Global Compact's Impact* (2004), at 4.

Figure 7.1 Impact of global compact on company reform

mean that different ones associate with the same norm in the same way. In other words, we are no better off knowing what function or role they play within the Global Compact itself because of the variety of different ways that companies can interpret them. Instead of assessing the significance of the principles in terms of how well actors comply with them or feel obligated to change their behaviour based on their normative pull, this chapter, based on the general approach of this work, takes a different perspective on such issues.[14]

---

corporations and stakeholders without taking into account other international initiatives. See also William Meyer and Boyka Stefanova, 'Human Rights, the UN Global Compact, and Global Governance' (2001) 34 *Cornell International Law Journal* 501, 504 for a discussion of the abstract nature of the human rights provisions of the GC in the context of other measures internationally.

[14] The ability of the GC to encourage compliance with its principles has already been studied by others. Importantly see Viljam Engström, *Realizing*

This chapter examines whether the GC framework, coordinated through the GCO, is potentially changing and instantiating a collective culture globally, whereby environmentally conscious approaches to consumption and production are the way that corporations will view their relationship to nature.[15] This is not a normative argument about whether the initiative is good or bad for the environment, but rather an expression of the kind of deeper cultural change in international politics that this initiative is potentially creating if it continues to develop.[16] It is argued that in this sense the role and function of environmental principles within the GC are better assessed in terms of how they frame ideas while participants and stakeholders interact within the regime. This interplay between environmental principles and the social influence of ideas is an important steering mechanism for the kind of learning that is potentially taking place within the GC. It also distinguishes this discussion from other attempts to consider the role of internationally developed voluntary codes as a common frame for multinational corporations to self-regulate.[17]

Initially this chapter describes the GC, highlighting the way that some of the initiatives by the GCO give structure to the engagement of corporate participants and stakeholders. The GC was selected for this study because through it transnational and other kinds of corporations directly

---

the Global Compact (2002); Meyer and Stefanova, above n 13; Elisa Morgera, 'The UN and Corporate Environmental Responsibility: Between International Regulation and Partnerships' (2006) 15 *Review of European Community and International Environmental Law* 93. Gjølberg and Ruud, above n 6, 12, have also observed, based on interviews with some Norwegian members of the GC, that the companies felt they would 'benefit from the legitimacy of the UN/GC while the voluntary nature and the abstract ten principles of the GC would 'make it hard to evaluate compliance.'

[15] For a different consideration of this question in the broader context of sustainable governance see Surya Deva, 'Sustainable Good Governance and Corporations: An Analysis of Asymmetries' (2006) 18 *Georgetown International Environmental Law Review* 707.

[16] For other attempts to situate the work of the GC in international relations topics see Meyer and Stefanova, above n 13. See also Bigge, above n 6, for an attempt to situate the GC within social constructivist work in international relations.

[17] On corporate responsibility through codes of conduct see Sean Murphy, 'Taking Multinational Corporate Codes of Conduct to the Next Level' (2005) 43 *Columbia Journal of Transnational Law* 389; Elisa Westfield, 'Globalization, Governance, and Multinational Enterprise Responsibility: Corporate Codes of Conduct in the 21st Century' (2002) 42 *Vanderbilt Journal of International Law* 1075; Ilias Bantekas, 'Corporate Social Responsibility in International Law' (2004) 22 *Boston University International Law Journal* 309; Deborah Leipziger, *The Corporate Responsibility Code Book* (2003).

and diffusely engage with each other and stakeholders at the international level using environmental principles. Additionally the GC embeds the principles within an institutional structure that relies more on social influence than coercive mechanisms that emphasise immediate gains. In section 7.2. the environmental principles of the GC are also examined as abstract and open-textured norms. Section 7.3 which follows argues that collective learning is in fact possible for multinational corporations at the international level. It examines what it means for them to instantiate a collective culture of responding in an environmentally responsible way to consumerism and production. Sections 7.4 and 7.5 discuss two different social processes facilitated by the GC to influence corporations to act in a way that will instantiate the collective learning of a culture of stronger environmental stewardship by corporations. This chapter concludes by discussing how environmental principles through social influence can help instantiate a collectively understood culture which moves away from excessive consumerism and production. This type of change deeply influences the nature of international cooperation amongst states and corporations and normative developments in the system.

7.2 THE GLOBAL COMPACT, THE UNITED NATIONS AND ENVIRONMENTAL PRINCIPLES

As far back as 1974 the UN through the Commission on Transnational Corporations took initiatives to develop a Code of Conduct for transnational corporations to establish a legal framework of some form to regulate what they do.[18] It has been argued that the 'UN's attempt to regulate transnational companies through its Code of Conduct produced 20 years

---

[18] See United Nations Department of Economic and Social Affairs, *The Impact of Multinational Corporations on Development and on International Relations* (1974). The Commission on Transnational Corporations was asked by UN ECOSOC to 'evolve' a 'set of recommendations, which, taken together, would represent a code of conduct for governments and TNCs to be considered and adopted by the Council' (ibid, 52). On this topic see in particular Barbara Frey, 'The Legal and Ethical Responsibilities of Transnational Corporations in the Protection of International Human Rights' (1977) 6 *Minnesota Journal of Global Trade* 153; Sidney Dell, 'The United Nations Code of Conduct on Transnational Corporations' in Johan Kaufmann (ed), *Effective Negotiation: Case Studies in Conference Diplomacy* (1989) 53; Peter Muchlinski, 'Attempts to Extend the Accountability of Transnational Corporations: The Role of UNCTAD' in Menno Kamminga and Saman Zia-Zarifi (eds), *Liability of Multinational Corporations Under International Law* (2000) 97.

of debate and negotiations, but yielded no results'.[19] The failed attempts to agree on how to manage corporations is seen as the reason why there was an absence in the international political economy literature of discussions of codes of conduct for transnational corporations of any sort between the 1980s and 1990s.[20] Only in the late 1990s, with the resurgence of discussions on corporate responsibility built round the *Norms on the Responsibilities of Transnational Corporations and Other Business Enterprises with Regard to Human Rights,*[21] did UN bodies develop a renewed interest in international codes of conduct. The GC initiative needs to be viewed in this context, especially because through the ten principles mentioned earlier in section 7.1, it adopts a softer approach to regulation. The overall mission and objectives of the GC are defined broadly:

> to be the world's most inclusive voluntary initiative to promote responsible corporate citizenship, ensuring that business, in partnership with other societal actors, plays its essential part in achieving the United Nations' vision of a more sustainable and equitable global economy.[22]

Two objectives which the GCO seeks to pursue in giving shape to this mission are stated as:

---

[19] Georg Kell, 'The Global Compact, Selected Experiences and Reflections' (2005) 59 *Journal of Business Ethics* 69, 73.

[20] Kathryn Sikkink, 'Codes of conduct for transnational corporations: the case of the WHO/UNICEF code' (1986) 40 *International Organization* 815, 815.

[21] See the 2003 revised version: *Norms on the Responsibilities of Transnational Corporations and Other Business Enterprises with Regard to Human Rights,* UN ESCOR, Commission on Human Rights, Sub-Commission on the Promotion and Protection of Human Rights, 55th sess, Agenda Item 4 UN Doc E/CN.4/Sub.2/2003/12/Rev.2 (2003), http//globalpolicy.org/socecon/tncs/2003/08ecosocnorms.pdf ('*UN Norms on the Responsibilities of TNCs*'). Although this initiative has so far failed, the ideas contained within the *UN Norms on the Responsibilities of TNCs* contributed, for instance, to a consultation paper that the Office of the High Commissioner for Human Rights and the GCO produced in conjunction with the 2005 Business Leaders Initiative on Human Rights: Business Leaders Initiative on Human Rights, United Nations Global Compact, and Office of the High Commissioner for Human Rights, *A Guide for Integrating Human Rights Into Business Management* (2005) United Nations Global Compact, www.unglobalcompact.org/docs/issues_doc/human_rights/guide_hr.pdf. For a discussion of the *UN Norms on the Responsibilities of TNCs* in terms of the role of the UN and states in developing it see Larry Backer, 'Multinational Corporations, Transnational Law: The United Nations' Norms on the Responsibilities of Transnational Corporations as a Harbinger of Corporate Social Responsibility in International Law' (2006) 37 *Columbia Human Rights Law Review* 287.

[22] *Letter to Global Compact Stakeholder*, above n 1, 2.

- Making the Compact and its principles on human rights, labour, environment and anti-corruption an integral part of business operations and activities everywhere;
- Encouraging and facilitating dialogue and partnerships among key stakeholders in support of the ten principles and broader UN goals, such as the Millennium Development Goals.[23]

The mission statement and the objectives of the GC confirm the fact that it seeks to do more than just identify ten principles that corporations must comply with. Although at first glance the GC initiatives appear as a code of conduct or a 'third party' 'principled code',[24] the broader governance framework,[25] which is coordinated by the GCO, is also relevant for differentiating this initiative from others at the international level, such as the *Guidelines for Multinational Enterprises*.[26] This is important because the GCO seeks to involve corporations in its work in direct and diffuse ways rather than assume that they are passive recipients of an international code of practice. It is therefore this broader framework established by the GC which engages and manages the external relations of corporations with other participants, stakeholders and the core ten principles as the pivot around which the governance structure of the GC is built.

The GC has sought to develop from the beginning as an initiative with a variety of different participants and stakeholders. It brings transnational and other kinds of corporations together with United Nations agencies, labour, civil society organisations and governments in an effort, to use the words of the GCO itself, 'to advance universal environmental and social principles in order to foster a more sustainable and inclusive world economy'.[27] Corporations are referred to as participants in the GC, and in

---

[23] Ibid.

[24] See, for instance, Christopher Wright and Alexis Rwabizambuga, 'Institutional Pressures, Corporate Reputation, and Voluntary Codes of Conduct: An Examination of the Equator Principles' (2006) 111(1) *Business and Society Review* 89, 93.

[25] The term 'governance framework' was more recently coined by the GCO to describe the range of activities that it coordinates to 'foster greater involvement in, and ownership of, the initiative by participants and stakeholders': Global Compact Office, above n 7, 16.

[26] The Organisation for Economic Cooperation and Development, *The OECD Guidelines for Multinational Enterprises: Text, Commentary and Clarifications*, OECD Doc.DAFFE/IME/WPG(2000)15/FINAL (2001) ('*OECD Guidelines*').

[27] Global Compact Office, *The United Nations Global Compact: Advancing Corporate Citizenship* (2005) United Nations Global Compact 1, www.unglobalcompact.org/docs/about_the_gc/2.0.2.pdf.

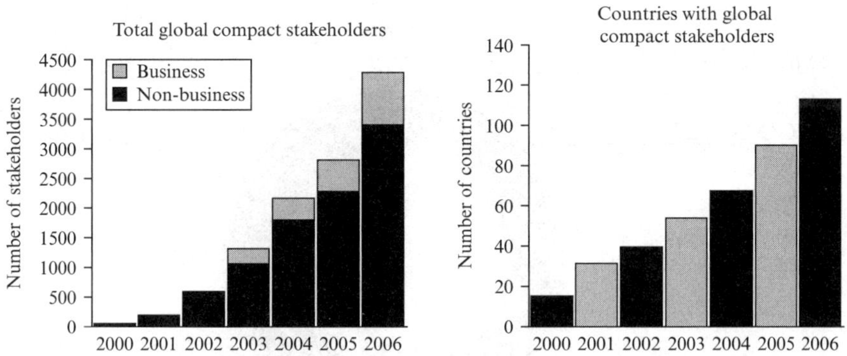

Source: Global Compact Office, *UN Global Compact Annual Review 2007 Leaders' Summit* (2007), 7.

Figure 7.2 Stakeholders in the global compact

2007 over 3000 from 116 countries were members of the GC.[28] The term stakeholder refers to anyone else who is a member of the GC. Figure 7.2. represents the increasing percentage of non-business stakeholders in the GC initiative. It also suggests that the multi-stakeholder approach has developed across the globe, given that the increase in both business and non-business participation matches the increase in the number of countries involved in the GC. However, in a useful critique of this development it has been suggested that there is a regional imbalance in both participants and stakeholders in the GC.[29] Figure 7.3. represents the diversity of the stakeholders other than corporations that have engaged with the GC initiative. This suggests that the engagement of non-business stakeholders with the GC is very diverse and potentially strong as a result of it.

The GCO spearheads the GC. It is located within the executive office of the Secretary-General of the UN because of the way it networks itself through various 'core agencies' of the UN.[30] The placement of the GC within the UN means that it is ultimately backed by states. In fact Kell identifies some governments as playing a very strong role in the formation

28 Global Compact Office, above n 7, 7.

29 Deva, above n 2, 134–136.

30 These include the Office of the High Commission for Human Rights (OHCHR), the International Labour Organisation (ILO), the UN Environment Programme (UNEP), the UN Development Programme (UNDP), the UN Industrial Development Organisation (UNIDO), and the United Nations Office on Drugs and Crime (UNODC): ibid, 15; Kell, above n 19, 75.

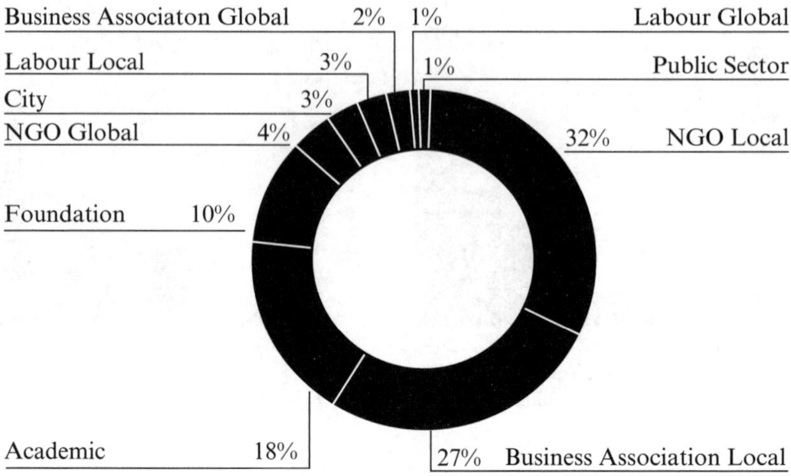

Source: Global Compact Office, *UN Global Compact Annual Review 2007 Leaders' Summit* (2007), 15.

Figure 7.3 Non-business participants by type

of the GC.[31] Their continuing involvement with the GC has been characterised as auxiliary in nature 'through outreach support, advocacy, and funding'.[32] In an official communication from the GCO, the role of states is also described as giving the 'essential legitimacy and universality to the principles of the Compact'.[33]

The GCO has identified three ways that it seeks to engage with corporate participants, the range of stakeholders represented in Figure 7.3., agencies of the UN, and the ten core principles of the GC.[34] They include: (1) getting commitment from the leadership of companies which are participants in the GC; (2) developing and implementing policies relating to the ten core principles and ensuring that the relevant people at the grassroots engage with them; and (3) ensuring that participants communicate their progress

[31] Ibid, 74. He noted that the United Kingdom, Sweden, Norway and Germany provided practical and financial support in forming and initially developing the GC.

[32] Georg Kell and David Levin, 'The Global Compact Network: An Historic Experiment in Learning and Action' (2003) 108(2) *Business and Society Review* 151, 153.

[33] Global Compact Office, above n 27, 4.

[34] Global Compact Office, above n 7, 9.

in relation to the principles.[35] The GCO has a variety of established practices that seeks to enhance the engagement of the different groups with the core principles. To help develop and implement policies, two strategies supported by the GCO stand out. The first is the establishment of the GC Local Networks (GCLN) which are defined as 'clusters of participants who come together to advance the Global Compact and its principles at the local level'.[36] The GCO has managed to establish a very large number of GCLNs to support its work within countries around the world.[37] Since 2000 when the first four GCLNs were established they have now multiplied to over 61 with apparently another 24 in development.[38] In an overview of GCLNs the GCO has said that they are meant to be 'moving innovative solutions upstream for global replication and multiplication', or 'taking global dialogue issues down to the level of implementation'.[39] This means that GCLNs help companies to implement the ten core principles and to facilitate their reporting obligations as participants of the GC.[40]

The second is the idea that the GCO supports the learning process for all involved through a variety of activities like conferences and workshops which seek to enhance the understanding of participants and stakeholders in relation to the core principles. Through these activities the GCO has for instance published a number of case studies and analyses to enhance the engagement of different groups.[41] Other UN agencies have also worked with stakeholders to develop materials to support the learning environment for corporations. For instance, in 2005 the UNDP along with the Danish Ministry of Foreign Affairs produced a 'Practical Guide for Companies Operating in Developing Countries'.[42] The guide aimed to assist companies with implementing the principles into their business

---

35 Ibid.

36 Global Compact Office and the Barcelona Centre for the Support of the Global Compact, above n 2, 8.

37 Ibid, 15.

38 Ibid.

39 Global Compact Office, *Networks Around the World*, United Nations Global Compact, www.unglobalcompact.org/NetworksAroundTheWorld/index.html.

40 Global Compact Office and the Barcelona Centre for the Support of the Global Compact, above n 2.

41 For instance see Global Compact Office, *Experiences in Management for Sustainability* (2003) United Nations Global Compact, www.unglobalcompact.org/docs/news_events/8.1/exp_man.pdf; Global Compact Office, *From Principles to Practice* (2003) United Nations Global Compact, www.unglobalcompact.org/docs/news_events/8.1/princ_prac.pdf.

42 KPMG, Royal Danish Ministry of Foreign Affairs and the United Nations Development Programme, *Implementing the UN Global Compact: A Booklet for*

operations in developing countries. Although the initiatives in this area appear very strong, whether or not corporations actually engage with the learning process is yet to be tested empirically.[43]

An important dimension of engaging corporations with other participants and stakeholders is the requirement that all corporations produce an annual Communication on Progress (COP), which they must share with stakeholders and also post on the official website of the GC.[44] As a voluntary and non-legally binding arrangement the reporting process of the GC is the only compulsory aspect of membership for corporations. To remain an 'active member' of the GC a corporation must submit its COP annually.[45] A COP must contain a statement that shows the continued support of the corporation for the core principles; describes what actions it has taken to implement the principles and whether it has engaged in partnership projects that support the goals of the UN more generally; and indicate how it has succeeded in meeting its goals using indicators or metrics available to it.[46] The GCO has reported that it has made 600 companies 'inactive' for not having submitted their COPs in the manner required by the GCO.[47] The main reason given for deactivating the membership of companies which do not report is to maintain the integrity of the initiative.[48] June 2005 was the first time that all corporations which had been members of the GC for more than two years had to report officially. The GCO reported that 87 per cent of the 102 Fortune 500 companies had reported.[49] However, only 25 per cent and 11 per cent of the medium and

---

Inspiration (2005), www.unglobalcompact.org/docs/news_events/8.1/dk_book_e.pdf.

[43] As an early example of an effort to do this see, for instance, Gjølberg and Ruud, above n 6, 11.

[44] Global Compact Office, *Policy for 'Communication on Progress'* (2008) United Nations Global Compact 2–3, www.unglobalcompact.org/docs/communication_on_progress/COP_Policy.pdf. See also Global Compact Office, *Leading the Way in Communication on Progress* (2006) United Nations Global Compact, www.unglobalcompact.org/docs/communication_on_progress/4.3/leading_the_way.pdf.

[45] This requirement kicks in after two years of joining the GC: Global Compact Office, above n 7, 51.

[46] Global Compact Office, *Policy for 'Communication on Progress'*, above n 44, 1.

[47] Global Compact Office, above n 7, 51.

[48] Global Compact Office, *Policy for 'Communication on Progress'*, above n 44, 2–3.

[49] Global Compact Office, *The Global Compact Communication on Progress: A Status Report and Recommendations for Improvements* (2005) United Nations Global Compact 5, www.unglobalcompact.org/docs/news_events/9.1_news_archives/2005_07_15/statrep_cop2.pdf.

small companies respectively had reported in 2005.[50] The reporting carried out by the larger companies is impressive, to say the least. What it says about smaller and medium-sized companies is that the benefits of membership do not warrant the need to support the GC's initiatives. Alternatively, the costs of reporting could be high. Whatever the reason, it appears that the types of corporations reporting are predominantly the larger transnational corporations which probably benefit from the kind of citizenship that is established through the GC.

7.2.1 Environmental Principles of the Global Compact

In 2000 the Trade Committee of the Organisation for Economic Cooperation and Development (OECD) compared 246 voluntary codes of conduct that corporations of its 29 member states could adopt if they wanted to.[51] They found that 'environmental stewardship' was 'one of the most heavily cited of the areas in the extended inventory'.[52] They noted that 145 out of the 246 codes dealt with environmental issues and 24 were exclusively dedicated to it.[53] However, given that a majority of these 246 codes discussed in the *Codes of Corporate Conduct Study* are developed by companies themselves or associations of various kinds, it is fair to presume that they deal with either product or process standards.[54] Research suggests that directly regulating particular activities of transnational corporations is achieved more effectively through 'narrowly defined issues' within codes of conduct.[55] This suggests that although there is nothing novel about another voluntary code, the fact that the GC adopts just three environmental principles to apply to corporations is somewhat unique.

---

50 Ibid.

51 Trade Committee and the Committee on International Investment and Multinational Enterprises of the OECD, 'Codes of Corporate Conduct – An Expanded Review of their Contents' (Working Paper 2001/6, Organisation for Economic Co-operation and Development, 2001), www.oecd.org/dataoecd/57/24/1922656.pdf ('*EOCD Codes of Corporate Conduct Study*').

52 Ibid, 11.

53 Ibid.

54 The distinction between product and process standards is from Philipp Pattberg, 'The Influence of Global Business Regulation: Beyond Good Corporate Conduct' (2006) 111 *Business and Society Review* 241, 244. The *EOCD Codes of Corporate Conduct Study*, above n 54, at 8 identified that 48 per cent of the voluntary codes were developed by companies; 37 per cent through associations; 13 per cent as partnerships of stakeholders; and 2 per cent by international organisations.

55 Sikkink, above n 20.

Much has been made of the inability of the general and short statements that make up the principles of the GC to provide a detailed and potentially useful frame for regulating corporate activities.[56] This is the case even amongst participants in the GC who have commented on the fact that the brevity of the principles does not do much in terms of regulating what they do.[57] The principles of the GC have been referred to by those involved in its design as aspirations[58] or shared values.[59] The GC's documents themselves refer to the principles as a value-based platform.[60] Whether or not they are aspirations or values of transnational or local corporations is an empirical question that does not seem as important as identifying how, as abstract and open-textured norms, they might function within the GC framework.

The three principles as abstract and open-textured norms rely on international environment law, defined broadly, for their meaning. Based on the origin of the three environmental principles it is apparent that the GC constructed these principles in an abstract and open-textured way. The three principles are drawn from the 1992 *United Nations Declaration on Environment and Development*[61] and *Agenda 21*,[62] which were developed for the United Nations Conference on Environment and Development (UNCED).[63] As a result, in applying environmental

---

56 See, for instance, Murphy, above n 17, 425; Klaus Leisinger, 'Opportunities and Risks of the United Nations Global Compact' (2003) 11 *Journal of Corporate Citizenship* 113; Deva, above n 2, 129–133.

57 See for instance the interviews conducted by Gjølberg and Ruud, above n 6, 14–16.

58 John Ruggie, 'Reconstituting the Global Public Domain – Issues, Actors, and Practices' (2004) 10(4) *European Journal of International Relations* 499, 516.

59 Kell and Levin, above n 32, 152.

60 See United Nations Global Compact, above n 9.

61 UN Doc A/CONF.151/5/Rev.1 (1992) ('*Rio Declaration*'). For commentary on the *Rio Declaration* see, for instance, David Wirth, 'The Rio Declaration on Environment and Development: Two Steps Forward and One Back, or Vice Versa' (1995) 29 *Georgia Law Review* 599; Ileana Porras, 'The Rio Declaration: A New Basis for International Cooperation' in Philippe Sands (ed), *Greening International Law* (1993) 20; Marc Pallemaerts, 'International Environmental Law From Stockholm to Rio: Back to the Future?' in Philippe Sands (ed), *Greening International Law* (1993) 1.

62 A copy of *Agenda 21* is available at www.un.org/esa/sustdev/documents/agenda21. *Agenda 21* was designed to clarify the scope of the environmental principles in the *Rio Declaration* but also contained principles in its own right: Nicolas de Sadeleer, *Environmental Principles: From Political Slogans to Legal Rules* (2005) 312.

63 See United Nations Global Compact, 'The Ten Principles – Environment', www.unglobalcompact.org/AboutTheGC/TheTenPrinciples/environment.html,

principles to corporations the GC uses ideas from within the broader international environmental legal framework that states have been responsible for developing since 1992.

The differences between the three principles are noteworthy. Principle 7 of the GC requires that 'business should support a precautionary approach to environmental challenges'. The fact that the GC adopted the term 'precautionary approach' from Principle 15 of the *Rio Declaration*, instead of the term precautionary principle which is also commonly used in the general literature[64] suggests that it sought to draw on the legitimacy of the document for the origin of the norm.[65] The more detailed definition of the precautionary approach in Principle 15 of the *Rio Declaration* is also used in the GC documents to support the abbreviated way in which it is stated as Principle 7 of the GC itself.[66] Put in this way, the precautionary approach appears more as a heuristic device which is stated in an abstract way as one of the GC principles. As a heuristic device it refers to an

---

for the environmental principles and brief description of their origin. See David Freestone, 'The Road from Rio: International Environmental Law after the Earth Summit' (1994) 6 *Journal of Environmental Law* 193 about the impact of the United Nations Conference on Environment and Development from the perspective of environmental norms.

64 Importantly Art 14 of the 2003 revised version of the *Norms on the Responsibilities of Transnational Corporations and Other Business Enterprises with Regard to Human Rights* adopts the term 'precautionary principle': Commission on Human Rights, Sub-Commission on the Promotion and Protection of Human Rights, *Norms on the Responsibilities of Transnational Corporations and Other Business Enterprises with Regard to Human Rights,* UN ESCOR, 55th sess, Agenda Item 4 UN Doc E/CN.4/Sub.2/2003/12/Rev.2 (2003).

65 There are a variety of different views on the difference between a precautionary principle and an approach: see Jacqueline Peel, 'Precaution – A Matter of Principle, Approach or Process?' (2004) 5 *Melbourne Journal of International Law* 483. As to the concept of the precautionary principle see Arie Trouwborst, 'The Precautionary Principle in General International Law: Combating the Babylonian Confusion' (2007) 16 *Review of European Community & International Environmental Law* 185; Arie Trouwborst, *Precautionary Rights and Duties of States* (2006).

66 Principle 15 of the *Rio Declaration* states that:

> [i]n order to protect the environment, the precautionary approach shall be widely applied by States according to their capabilities. Where there are threats of serious or irreversible damage, lack of full scientific certainty shall not be used as a reason for postponing cost-effective measures to prevent environmental degradation.

This exact definition is also adopted by the GC as the definition of the precautionary approach for businesses: see United Nations Global Compact, 'The Ten Principles – Principle 7', www.unglobalcompact.org/AboutTheGC/TheTenPrinciples/principle7.html.

articulation of an established set of ideas which are stated in more open-textured fashion in Principle 15 of the *Rio Declaration*.

Principle 8 of the GC requires that 'businesses should undertake initiatives to promote greater environmental responsibility'. Unlike the precautionary approach, this principle is extracted from Chapter 30 of Agenda 21, which is more generally dedicated to the role of businesses and industry.[67] The GC draws from two different places to develop the potential intersubjectivity around the vague idea of assuming greater environmental responsibility.[68] First, the GC refers to Chapter 30.26 of Agenda 21 which requires that transnational corporations:

> should ensure responsible and ethical management of products and processes from the point of view of health, safety and environmental aspects. Towards this end, business and industry should increase self-regulation, guided by appropriate codes, charters and initiatives integrated into all elements of business planning and decision-making, and fostering openness and dialogue with employees and the public.

Secondly, the documents of the GC refer to Principle 2 of the *Rio Declaration* which requires that states take 'responsibility to ensure that activities within their jurisdiction or control do not cause damage to the environment of other States or of areas beyond the limits of national jurisdiction'. The combination of these two very different expectations means that Principle 8 imposes an internal obligation on corporations to self-regulate and an external one to consider the impact they will have on others. Principle 8 is still very open-textured and vague despite the connections that the GC tries to develop between it and international environmental soft law more generally. This is because, unlike the precautionary principle for instance, the way that Principle 8 is linked to the *Rio Declaration* and *Agenda 21* is not obvious. As a result the lack of an apparent intersubjective framework means that the vagueness inherent in the concepts in Principle 8 has different functional implications for social learning from Principle 7.

Principle 9 requires that 'business should encourage the development and diffusion of environmentally friendly technologies'. Given the reliance of the other two principles on Agenda 21 and the *Rio Declaration* this principle appears to have been directly adapted from Principle 9 of the *Rio Declaration* which requires that:

---

[67] See United Nations Global Compact, 'The Ten Principles – Principle 8', www.unglobalcompact.org/AboutTheGC/TheTenPrinciples/principle8.html.

[68] Ibid.

> States should cooperate to strengthen endogenous capacity-building for sustainable development by improving scientific understanding through exchanges of scientific and technological knowledge, and by enhancing the development, adaptation, diffusion and transfer of technologies, including new and innovative technologies.

The idea that states should adopt the 'best available technology' or 'best practical means' for doing something is a common requirement of many multilateral agreements, and therefore more could have been added to the provisions of these principles had the GC chosen to do so.[69] Compared to Principles 7 and 8 this principle is arguably more defined but still remains open-textured in the direction it gives to actors.

This discussion highlights the possibility that these environmental principles which have developed out of qualitative concepts like 'precaution' or 'responsibility' can encourage a range of actors with a variety of beliefs to internalise their meaning for themselves. Actors have different values and beliefs behind why they might protect and preserve the natural environment and resources. Not only do these values and beliefs elicit different responses from actors, but also the depth of reaction to the same problem can also vary significantly.[70] For instance, a state may preserve a rainforest for its biodiversity and because of the aesthetic, spiritual, historical or symbolic value that it has for all or particular parts of its current or future population. The diversity of values in the context of a regime like the GC where corporations and other stakeholders like governments, civil society and labour organisations are involved in different ways is bound to be significant.

Additionally the GC has generated principles in terms that presume

---

[69] For instance the *Convention on Long-Range Transboundary Air Pollution*, opened for signature 13 November 1979, 1302 UNTS 218 (entered into force 16 March 1983 Art 6 requires that states adopt the 'best available technology'. The *United Nations Convention on the Law of the Sea*, opened for signature 10 December 1982, 1833 UNTS 397 (entered into force 16 November 1994), Art 194 requires that states adopt what is practically the best available means at their disposal to prevent, reduce and control pollution of the marine environment. Other recent uses and references to the principles in water pollution agreements include the *Convention of the Protection and Use of Transboundary Watercourses and International Lakes*, opened for signature 17 March 1992, 31 ILM 1312 (entered into force 6 October 1996) and the *Convention for the Protection of the Marine Environment of the North-East Atlantic*, opened for signature 22 September 1992, 31 ILM 1069 (entered into force on 25 March 1998) ('*OSPAR Convention*').

[70] This point has been used for criticising the use of abstract principles within the GC framework. See Deva, above n 2, 129–133.

their relevance for what corporations might value. This suggests that identifying what relevance the principles have for corporations is in itself a function of the GC governance framework rather than something to be presumed. This means that it is unproductive to assess the extent to which corporations should comply with the environmental principles of the GC. In fact, Kell and Ruggie, as two main architects and drivers of the GC, have noted that it was not designed as a code of conduct. They write that '[i]nstead, it is meant to serve as a framework of reference and dialogue to stimulate best practices and to bring about convergence in corporate practices around universally shared values'.[71] In other words the participation of corporations is important for creating meaning from the environmental principles used in the GC and it cannot be presumed that they are values that corporations naturally relate to.

The relevance of the abstract and open-textured style of Principles 7, 8 and 9 of the GC as discussed can also be contextualised by comparing them to other voluntary codes developed internationally for transnational corporations. For instance, the *OCED Guidelines* also provides recommendations on how international business should conduct itself in relation to a range of areas including the protection of the environment.[72] In comparison, for instance, the *OECD Guidelines* do not use an abstract abbreviation of the precautionary principle, but instead define it in full and in a way that has subtle and important differences from Principle 15 of the *Rio Declaration*.[73] In other words the *OECD Guidelines* seek to specify in a more precise way how the precautionary principle has to be interpreted by the international business community. This does not appear to have effectively provided for the application of the precautionary principle to corporations, as is apparent for instance in the issues raised during the 2004 Annual Meeting which reviewed the *OECD Guidelines*.[74] This is because

---

71 Georg Kell and John Ruggie, 'Global Markets and Social Legitimacy: the Case for the "Global Compact"' (1999) 8(3) *Transnational Corporations* 101, 104.

72 In particular, ch 5 of the *OECD Guidelines* provides eight different fairly detailed recommendations for international business: OECD, *The OECD Guidelines for Multinational Enterprises: Text, Commentary and Clarifications*, OECD Doc.DAFFE/IME/WPG(2000)15/FINAL (2001) 28–29.

73 Ibid, ch 5(4) provides that:

> [c]onsistent with the scientific and technical understanding of the risks, where there are threats of serious damage to the environment, taking also into account human health and safety, not use the lack of full scientific certainty as a reason for postponing cost-effective measures to prevent or minimise such damage.

74 OECD Directorate for Financial and Enterprise Affairs, *Roundtable on Corporate Responsibility: Encouraging the Positive Contribution of Business to*

certain concepts, like risk, are open-textured and potentially applicable to all sorts of instances of corporate activity. More importantly, what is an acceptable approach to risk was identified as important for the public sector to define, rather than individual corporations which might vary significantly in their approaches.[75] It appears from this that spelling out particular formulations of the ideas contained within abstractly defined environmental principles does not necessarily predispose them to being applied more easily by corporations.

The next section discusses the potential of corporations actually to share at the international level a culture that values anything in particular that is traditionally of concern to the public at large and states in particular. This argument seeks to position the environmental principles as norms which have significance for how corporations can learn collectively to identify with a particular culture.

7.3 COLLECTIVE LEARNING THROUGH THE GLOBAL COMPACT

An important criticism of the GC discussed above has been in terms of whether the objectives of the UN expressed through the principles and the GCO can be reconciled with the needs and preferences of private enterprise.[76] As an alternative, Ruggie has argued that, although the principles established as part of the GC were developed by states, their adoption by corporations suggests that they are also able to 'encompass the sphere

---

Environment Through the OCED Guidelines for Multinational Enterprises – Summary of the Roundtable Discussion (2004), www.oecd.org/dataoecd/54/24/33805553.pdf. For a discussion of the environmental provisions of the *OECD Guidelines* see Elisa Morgera, 'An Environmental Outlook on the OECD Guidelines for Multinational Enterprises: Comparative Advantage, Legitimacy, and Outstanding Questions in the Lead Up to the 2006 Review' (2006) 18 *Georgetown International Environmental Law Review* 751.

75 In the report of the 2004 meeting it was noted that:

> whereas the question of where to draw the line between government and corporate responsibility is an important cross-cutting issue, it is, particularly pertinent in the context of risk management. As a corporate representative said, few, if any activities are risk-less and so it is unrealistic to expect companies to shun all environmental risks. This is one reason why companies cannot be left alone managing risk; there needs to be a degree of involvement on the part of government and civil society to discuss *inter alia* what constitutes acceptable risk levels.

OECD Directorate for Financial and Enterprise Affairs, above n 74, 8.

76 See Engström, above n 14, 46–48.

of the transnational corporate activity'.[77] Part of the difficulty with this debate is the presumption that the principles are values and that the GC is being built on a value-based platform. As discussed above, environmental principles of the GC are open-textured or abstract norms which require actors to engage with them to create meaning from them.

The important question seems to be whether corporations are capable of 'genuine political activity' within the international order, whether within or '*apart from*' the system of states.[78] States engage in politics because they could be trying to solve cooperation problems by creating norms and transacting together. Ideational and normative structures develop, for instance, out of the need to solve issues that require the cooperation of actors, such as preventing the ozone layer's depletion. Whether the direct or diffused interactions of corporations with each other are capable of generating shared culture or intersubjectivity is an important issue. The potential of abstract and open norms like environmental principles to generate meaning for a group of corporations should not necessarily be presumed.

Traditionally the research agenda among scholars has focussed on the ability of corporations or civil society organisations to pressure states, intergovernmental negotiations and international agencies.[79] This research agenda is a reflection of a dominant and persistent ideology that still sees states as 'the primary form of political organization'.[80] There are different ways in which the role of transnational corporations and civil society organisations in relation to public matters like the environment is being rationalised within the international order. Wapner in 1995 developed the idea of the 'world civic politics' to argue that at the international level there was political activity apart from the state.[81] A second idea points out that there are 'deeper processes of globalization at work that

---

[77] Ruggie, above n 58, 516.

[78] Ibid, 502. See also Paul Wapner, 'Politics Beyond the State: Environmental Activism and World Civic Politics' (1995) 47(3) *World Politics* 311. Some have in fact turned these issues around and have investigated the potential role of private actors to compensate for the lack of initiative shown by states to fix or provide public goods: see Christoph Knill and Dirk Lehmkuhl, 'Private Actors and the State: Internationalization and Changing Patterns of Governance' (2002) 15 *Governance* 41.

[79] See, eg, Margaret Keck and Kathryn Sikkink, *Activists Beyond Borders: Advocacy Networks in International Politics* (1998).

[80] Ruggie, above n 58, 502.

[81] Wapner, above n 78. For considerations of this idea in the context of the GC see Ruggie, above n 58.

are producing a disengagement of law and State'.[82] This liberates other entities, including individuals and transnational corporations, to have a significant say in the processes of globalisation.[83] Importantly, Cutler has argued that the main problem with questions about the political activity of entities like transnational corporations is that 'conventional theories of international relations and international law are incapable of capturing' private authority.[84] She argues that the 'analytical, theoretical and ideological' orientations of these disciplines render such authority a '*non sequitur*'.[85] In the case of many approaches to international relations, states are still the primary actors responsible for change within the international order.[86] Public international law is even more explicitly state-centric, as it provides that only states can create law.[87] These ideas suggest the potential difficulty with transnational corporations shaping their own collective understanding of culture or norms relating to public issues.[88]

There is also the challenge of basing discussions of the political activities of corporations round norms. This is captured by Sol Picciotto, who highlights the 'complex and multi-layered interactions between laws, codes and guidelines, operating locally, nationally, transnationally, regionally and internationally'.[89] Santos also makes similar points in stating that '[o]ur life is constituted by an intersection of different legal orders, that is by interlegality. Interlegality is the key post-modern conception of law'.[90] These ideas suggest that even if we are able to observe change in corporate

---

82 A Claire Cutler, *Private Power and Global Authority: Transnational Merchant Law in the Global Political Economy* (2003) 32–3.

83 These ideas are also usefully captured in the important scholarship of Santos on transnational entities: Boaventura de Sousa Santos, *Toward a New Legal Common Sense* (2nd ed, 2002) 181–182.

84 Cutler, above n 82, 2.

85 Ibid.

86 See Ronen Palan, 'Recasting Political Authority: Globalization and the State' in Randall Germain (ed), *Globalization and Its Critics: Perspectives from Political Economy* (2000) 139, at 140.

87 For instance, the *Statute of the International Court of Justice* Art 38(1)(c) recognises the role only of civilised nations in creating general principles of law.

88 For a good survey and examination of the role of 'multinational enterprises' in governance see Lee Tavis, 'Corporate Governance and the Global Social Void' (2002) 35 *Vanderbilt Journal of Transnational Law* 487.

89 Sol Picciotto, 'Introduction: What Rules for the World Economy?' in Sol Picciotto and Ruth Mayne (eds), *Regulating International Business: Beyond Liberalization* (1999) 1, 17.

90 Boaventura de Sousa Santos, 'Law: A Map of Misreading. Toward a Post-modern Conception of Law' (1987) 14(3) *Journal of Law and Society* 279, 300.

behaviour towards a more socially responsible one, ascertaining its cause in a complex world is difficult, if not impossible. On the other hand, Cutler, in her scholarship on private authority, has argued that the mercatocracy actually 'functions to provide a unity of purpose and a coherence in regulation'.[91] She suggests that this is 'obscured by notions of pluralistic or fragmented governance'.[92] That is, concepts and ideas used in public international law make the mercatocracy appear more fragmented and pluralistic than it is.[93] The implications of Cutler's views are that the approaches of corporations to issues are not that seemingly fragmented to stop them from instantiating particular cultural ideals.[94] The issue that is of more concern is in terms of the indeterminate nature of how corporations come to prefer certain things, as highlighted by Santos and Sol Picciotto in the above references to their work, rather than whether they can collectively identify with something in particular.

7.3.1 Engagement and the Instantiation of a Collective Culture through the Global Compact

It is important to ask why corporations work together or participate in collectively organised initiatives. It is also necessary to inquire whether they can or will engage in political activities at the international level. In other words, why corporations wish to interact directly or diffusely through, for instance, the GC is also an important issue. In a 2007 study McKinsey & Company, who interviewed around 391 chief executive officers (CEOs) and top executives of corporations which were at that time participants in the GC, noted that:

> [m]ost businesses are facing overwhelmingly large ESG issues that span regions and industries, and threaten their long-term viability. More and more, businesses are collaborating to level the playing field – working in multi-sector partnerships with civil society organizations, governments, and each other.[95]

---

[91] Cutler, above n 82, 31 and ch 6.

[92] Ibid.

[93] Ibid.

[94] See also Peter Muchlinski, '"Global Bukowina" Examined: Viewing the Multinational Enterprise as a Transnational Law-Making Community' in Günther Teubner (ed), *Global Law Without a State* (1997) 79.

[95] McKinsey & Company, 'Shaping the New Rules of Competition: UN Global Compact Participant Mirror' (2007) 24, www.unglobalcompact.org/docs/summit2007/mckinsey_embargoed_until020707.pdf. The abbreviation ESG used in this quotation refers to 'environmental, social and governance'. In this same study, McKinsey & Company, at 7, report that 95 per cent of the 391 CEOs of

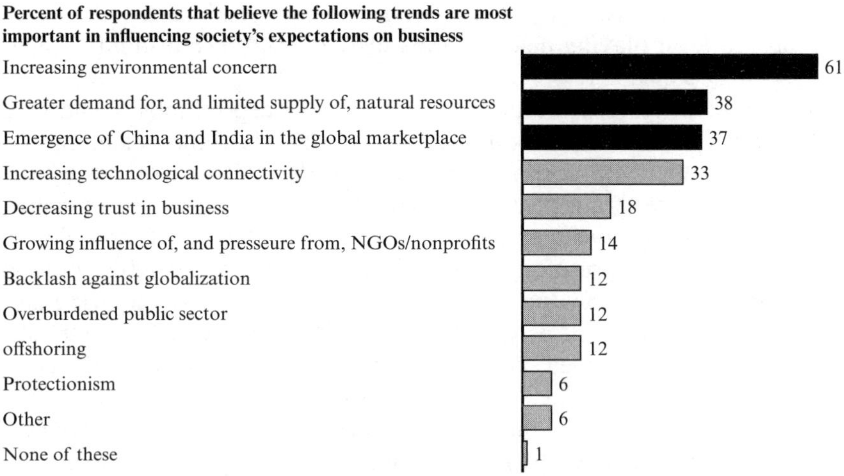

Source: Mckinsey & Company, 'Shaping the New Rules of Competition: UN Global Compact Participant Mirror' (2007), at 11.

Figure 7.4 Trends influencing society's expectations on business

This comment highlights for instance both the pressure to take responsibility for environmental stewardship and the need to level the playing field as two reasons for why corporations might collaborate through the GC. McKinsey & Company had also asked leaders within corporations to select 'developments linked to accelerated global patterns of growth as the three most important trends that are influencing society's expectations on businesses'.[96] From their responses, represented in Figure 7.4, corporations understand the severity of the pressure on them to deal with environmental concerns. In this same 2007 study McKinsey & Company also reported that many of the corporate leaders they surveyed indicated that they viewed the GC as 'one example of a coalition organized around a wider set of ESG issues'.[97] Arguably, the

companies they spoke to 'agreed that society has greater expectations for business to take on public responsibilities than it had 5 years ago'.

[96] Ibid, 10.

[97] Ibid, 24. There are lots of other ways to conceptualise why corporations would choose to join the GC. For instance, Gjølberg and Ruud, above n 6, at 11, have pointed out that during an interview with Norwegian companies they were told that the GC gives them 'a "social licence to operate". It is beneficial to the companies' reputation management, brand image, employee satisfaction, recruitment, stakeholder relations, customer satisfaction etc.'

GC is seen, amongst other things, as giving corporations the chance to establish a level playing field in terms of the social responsibility that is increasingly called for.

It is difficult to separate the real reason why corporations would seek to establish a collective culture of any kind.[98] From the beginning the UN partnership with corporations through the GC was driven by the ideology that free trade and open markets would be enhanced through more socially responsible behaviour.[99] That is, the GC sought to encourage how corporations made money rather than to affect growth by adopting restrictions on corporate behaviour.[100] This rhetoric plays itself out in a variety of ways within the work of the GC. For instance, the GCO has encouraged corporations to place importance on the cost-effectiveness of protecting the environment. In one of its guides to the GC it has suggested that the precautionary approach to doing business means avoiding damage to the environment because it is cheaper than remedying it in the future.[101] The GCO has noted that, in the long term, an investment into environmentally safer production methods saves money.

The culture that the GC is seeking to promote appears to encourage an approach to markets that is encapsulated by ideas developed within ecological economics.[102] According to Daly, who is argued to be amongst

---

[98] See Kell, above n 19, 74.

[99] Engström, above n 14, 11. Engström notes that this dimension of the GC has been criticised as giving in to the *realpolitik* of the balance of power internationally: ibid, 11 at fn 39. See in particular the following comment from the former UN Secretary General Kofi Annan in his statement at the launch of the GC in 1999, suggesting that the UN should 'help make the case for and maintain an environment which favours trade and open markets': United Nations, 'Secretary-General Proposes Global Compact on Human Rights, Labour, Environment, in address to World Economic Forum in Davos' (Press Release, 1 February 1999), www.un.org/News/Press/docs/1999/19990201.sgsm6881.html.

[100] Engström, above n 14, 11.

[101] Global Compact Office, *Guide to the Global Compact: A Practical Understanding of the Vision and Nine Principles* (2002) 54, www.aeca.es/comisiones/rsc/documentos_fundamentales_rsc/pactomundial/global_compact_guide.pdf.

[102] This discussion aims to illustrate the idea of a cultural shift rather than defend it. As a result the discussion of ecological economics is not particularly detailed. For more information see, for instance, Herman Daly, 'Sustainable Growth: An Impossibility Theorem' in Herman Daly and Kenneth Townsend (eds), *Valuing the Earth* (1993) 267; Joseph DesJardins, *Environmental Ethics: An Introduction to Environmental Philosophy* (4th ed, 2005). For an insightful and powerful critique see Mark Sagoff, *Price, Principle, and the Environment* (2004).

the pre-eminent scholars in this field, sustainable or ecological economics is based on distinguishing between the ideas of *development* and *growth*.[103] Daly has written that:

> To *grow* means to 'increase naturally in size by the addition of materials through assimilation or accretion.' To *develop* means 'to expand or realize the potentialities of; to bring gradually to a fuller, greater, or better State.' When something grows it gets bigger. When something develops it gets different. The earth ecosystem develops (evolves), but it does not grow. Its subsystem, the economy, must eventually stop growing, but can continue to develop. The term 'sustainable development' therefore makes sense for the economy, but only if it is understood as 'development' without growth.[104]

To illustrate the ideas of ecological economics one can make comparisons between different theoretical approaches to consumption and production. The first is the traditional market economics model which deals with the allocation of resources by giving priority to those who are willing to pay the most for it.[105] This means that a particular resource does not get developed and distributed unless there is consumer demand for a product that uses it.[106] In contrast, through ecological economics the use of resources is 'only at a rate that can be sustained over the long term and one that recycles or reuses both the by-products of the production process and the products themselves'.[107] Amongst other things, a culture driven by the ideas behind ecological economics would for instance not produce goods simply because there is consumer demand for them. The production of particular goods would be limited to the potential renewal of those natural resources needed for the process.[108]

An alternative ideological and cultural approach to ecological economics examines the lifestyle that society is seeking to sustain while we become more aware of the impact we have on the environment and resource use. It is critical of the idea of sustaining our current consumption and production patterns.[109] Sagoff, who criticises the emphasis on 'scarcity and overconsumption', adopts such an approach to ecological economics.[110]

---

103 See DesJardins, above n 102, 87 for this summary of Daly.
104 Daly, above n 102, 267–268.
105 DesJardins, above n 102, 87.
106 Ibid, 87–88.
107 Ibid, 88.
108 Ibid.
109 Ibid, 89.
110 See, for instance, Sagoff, above n 102; Mark Sagoff, *The Economy of the Earth: Philosophy, Law and the Environment* (1988).

His arguments are complex and a summary would not do justice to them. Sagoff's approach is a critique of the fact that we value nature as a resource and our understanding of consumption and production is therefore driven by conceptions of scarcity.[111] Instead, Sagoff emphasizes spiritual, aesthetic and ethical approaches to valuing goods, nature and the environment. These alternative approaches to ecological economics are presented here to illustrate the other end of the spectrum from the standard model of market economics.

Given current standards of living in the developed world, the idea that people in any society would accept restrictions on their consumer preferences is a significant departure from the norm. This suggests that the structural power of the kind of culture that ecological economists argue for is yet to dominate general consumption and production patterns around the world. Although there are going to be natural exceptions, the culture that traditional market economics defend is generally taken for granted. More importantly it is difficult to see how corporations will support an alternative culture to consumerism. In fact it was argued back in 1962 that our corporations were not designed or created to support social responsibility for protecting the environment.[112]

The point being made is that a corporation's preferences and wants vary significantly in comparison to those of other actors whose deeper values and beliefs are not shaped entirely by approaches to consumption and production. For corporations to adopt preferences in favour of environmental ideals that protect the future generation or restrain how they respond to potential consumer demands is in itself a normative shift in the way they behave. This is because corporations are more concerned with potential liabilities from harm and damage rather than protecting the environment by restraining production. It is therefore difficult to foresee how corporate behaviour and culture could change as dramatically to reflect the standards of ecological economics.

This discussion does not suggest that the shift in culture or the way that corporations view and approach their business is likely to be swift. The instantiation of a particular culture is in itself a collective learning experience. It is arguable that corporations are beginning to take steps to instantiate a collective culture globally where the approaches of ecological economics within business could become the norm rather than the exception. The GCO in 2007 completed a survey of 400 companies which illustrated that an important proportion of them have some kind of

---

111 This summary is taken from DesJardins, above n 102, 90.

112 Milton Friedman, *Capitalism and Freedom* (1962).

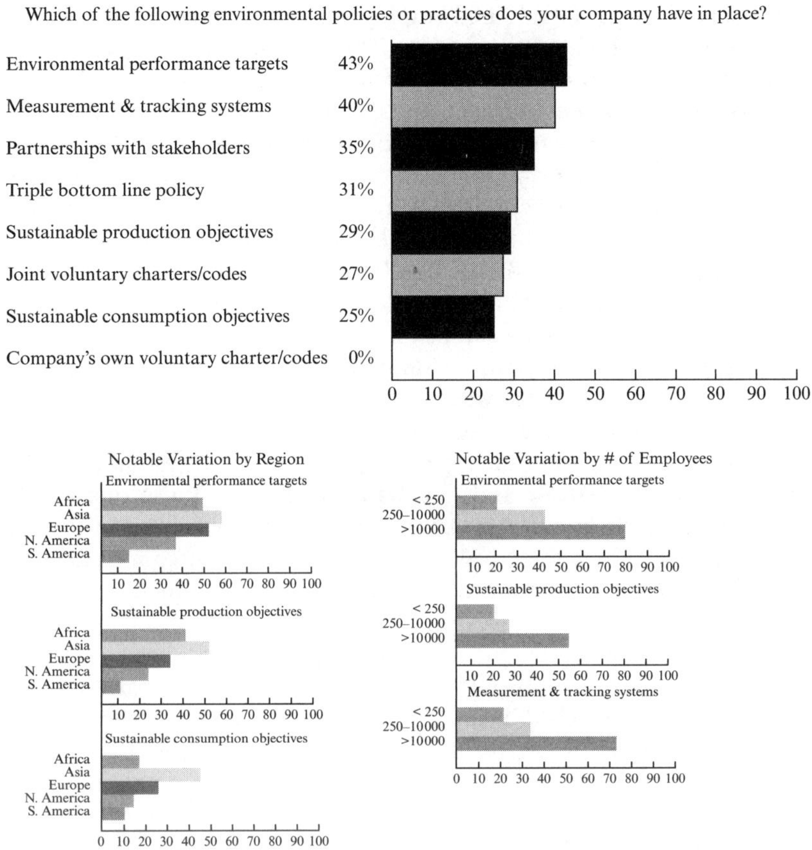

Source: Global Compact Office, *UN Global Compact Annual Review 2007 Leaders' Summit* (2007), 33.

Figure 7.5 Implementation of environmental policies and practices by corporations across countries and regions

environmental policies or practices in place.[113] The findings are presented in Figure 7.5.

According to the research presented in Figure 7.5 a significant proportion of those corporations which have adopted environmental policies or

113 Global Compact Office, above n 7, 33. Of the 400 companies that responded to the GCO survey, 28 per cent had more than 10,000 employees; 41 per cent had between 250–10,000 and 32 per cent had fewer than 250 employees: ibid, 9.

practices have more than 10,000 employees, which often means that they are large transnational corporations. This is important because of their 'capacity to locate production across national borders, to trade across frontiers, exploit foreign markets, and organize managerial structures in a way that affects the international allocation of resources'.[114] Also, because the production and distribution networks of transnational corporations usually span many jurisdictions they are more likely to have an influence on a variety of different stakeholders.[115] As a result a shift in the practices of transnational corporations is likely to be a much stronger instantiation of an emerging global culture driven by an ecological approach to issues of production and consumption.

In the 2007 McKinsey and Company report referred to earlier in relation to Figure 7.4,[116] it was noted that 'many CEOs recognize the underlying tension between business models wedded to increasing patterns of consumption, and the reality of limited natural resources'.[117] The findings of the McKinsey and Company report and those of the GCO discussed above strongly suggest that companies have started to rethink their approach to how they drive consumption of certain goods and their involvement in the extraction of natural resources for their production. This discussion does not suggest that the GC can of itself achieve the cultural shift of corporations taking ecological approaches to consumption and production. In fact, given the earlier references to the work of Santos and Sol Picciotto, it would be inaccurate to identify a likely cause for such a cultural shift in the patterns of growth. Despite this, the framework of the GC and the environmental principles within it can potentially shape and manage the interactions of corporations, particularly the 100 plus transnational corporations which are members of it, to instantiate or provide support for the development of such a cultural shift. For instance, the diffusion of ideas amongst corporations and the type of identity that the Communication on Progress (COP) policy of the GCO creates are potentially capable of establishing an important measure for instantiating a collective culture at the international level.

---

114 Engström, above n 14, 5. See also Peter Muchlinski, *Multinational Enterprises and the Law* (1995) 15.

115 Engström, above n 14, 5. See also Muchlinski, above n 114, 15. For discussions of the role of transnational corporations and global production see David Held, Anthony McGrew, David Goldblatt and Jonathan Perraton, *Global Transformations: Politics, Economics and Culture* (1999) especially ch 5.

116 McKinsey & Company, above n 95.

117 Ibid, 13.

7.4 DIFFUSION AND SOCIAL INFLUENCE IN THE CONTEXT OF THE GLOBAL COMPACT

Not all ideas relating to how corporations should behave are socially influential enough to be adopted or internalised. The scholarship on social learning and socialisation in international relations has developed a focus on the concept of diffusion because it explains one of the reasons why actors might internalise particular ideas or intersubjective structures like norms.[118] The diffusion of ideas, things, practices, or norms or policies is another way of referring to a process whereby innovations are 'communicated over time among members of a social system'.[119] Not all ideas that are diffused however are subsequently adopted or utilised in particular instances.[120] Becoming aware of an idea does not necessarily suggest that the actors will change how they behave.[121]

The idea that actors such as corporations would emulate or model themselves on what they see in others is a concept used in the scholarship

---

[118] Kai Alderson, 'Making Sense of State Socialization' (2001) 27 *Review of International Studies* 415, 425. On the idea that international organisations teach and diffuse norms see Martha Finnemore, 'International Organizations as Teachers of Norms: the United Nations Educational, Scientific, and Cultural Organization and Science Policy' (1993) 47 *International Organization* 565; Alexandru Grigorescu, 'European Institutions and Unsuccessful Norm Transmission: The Case of Transparency' (2002) 39(4) *International Politics* 467. For a more general review of diffusion in the context of international relations see Leslie Eliason and Emily Goldman, 'Introduction: Theoretical and Comparative Perspectives on Innovation and Diffusion' in Emily Goldman and Leslie Eliason (eds), *The Diffusion of Military Technology and Ideas* (2003) 1, especially 11–22.

[119] See Everett Rogers, *Diffusion of Innovations* (4th ed, 1995), cited in Helge Jorgens, 'Governance by Diffusion: Implementing Global Norms Through Cross-national Imitation and Learning' in William Lafferty (ed.), *Governance for Sustainable Development: The Challenges of Adapting Form to Function* (2004) 246, 252.

[120] This conceptual distinction is emphasised by Eliason and Goldman, above n 118, 12. Eliason and Goldman point to the following research as shaping this idea that they summarise in their work: Colin Bennett, 'How States Utilize Foreign Evidence' (1991) 11(1) *Journal of Public Policy* 31; Colin Bennett, 'What Is Policy Convergence and What Causes It?' (1991) 21 *British Journal of Political Science* 215; Colin Bennett, 'Understanding Ripple Effects: The Cross-National Adoption of Policy Instruments for Bureaucratic Accountability' (1997) 10(3) *Governance* 213.

[121] Bennett, 'How States Utilize Foreign Evidence', above n 120, 32–33. In particular, Bennett at 32–33 argues that there is a difference between '*knowledge* of a foreign program, *utilization* of that knowledge, and the *adoption* of the same program'.

on diffusion to describe how ideas spread across the globe.[122] The main rationale for emulating what another actor does is the importance that is often placed on ideas that are readily available and have been tried elsewhere in solving or dealing with similar problems or issues.[123] Braithwaite and Drahos distinguish modelling (or emulation) from imitation by suggesting that it is a process with cognitive content because ideas are communicated to others using words or images.[124] As a result emulation could mean that a policy will be used only to shift agendas rather than for actors to do exactly what others have done in the past.[125]

When actors like corporations exist in a competitive global business environment it is more likely that they will emulate or model themselves on others who are successful. Although Ikenberry in the following quotation is writing about states, the ideas share a degree of resonance with transnational and other corporations;

> [i]t appears that emulation will be most pervasive when international competition is most intense, inasmuch as competition provides a power incentive to monitor and respond to innovations developed abroad. In cases of intense competition, the costs of failing to innovate are highest. Moreover, the diffusion of policy practices is most likely to occur when states share underlying cultural or social settings (although a highly competitive environment might serve to transcend obstacles presented by divergent cultural or societal circumstances).[126]

In a competitive environment corporations are likely to be looking for ideas and innovations that might enhance their goals and objectives. Given that corporations generally aim to increase their annual profits a competitive environment would strongly favour emulating successful initiatives or modelling themselves to be like their peers.[127]

---

[122] On emulation and diffusion see, in particular, John Ikenberry, 'The International Spread of Privatization Policies: Inducements, Learning, and Policy Bandwagoning' in Ezra Suleiman and John Waterbury (eds), *The Political Economy of Public Sector Reform and Privatization* (1990) 99. On modelling and diffusion see John Braithwaite and Peter Drahos, *Global Business Regulation* (2000) especially ch 22.

[123] Kurt Weyland, 'The Diffusion of Innovations: A Theoretical Analysis' (Paper presented at the annual meeting of the American Political Science Association, Boston, 28 August 2002), cited in Jorgens, above n 119, 252.

[124] Braithwaite and Drahos, above n 122, 25.

[125] Bennett, 'What Is Policy Convergence and What Causes It?', above n 120, 221.

[126] Ikenberry, above n 122, 103, quoted in Eliason and Goldman, above n 118, 15–16.

[127] Braithwaite and Drahos, above n 122, ch 22.

This is not to suggest that competition and the pressure for corporations to emulate or model themselves on success has to be viewed in one-dimensional terms. The pressure to be more socially responsible in protecting the environment from harm and depletion of natural resources is also likely to influence corporations to emulate others who have done this successfully.[128] As a result companies feeling pressure from shareholders, consumers, distributors, or the general public would identify with other corporations which have performed strongly in taking responsibility towards environmental stewardship. This suggests that they are more likely to emulate or model themselves on others because their competitive position in the market could easily be impacted on in the long run if they did not take appropriate steps in this direction.

Modelling or emulating the activities of other actors is also about identifying with others rather than just making a rational choice to pursue a particular end.[129] This idea can be illustrated by reference to a study of Samarco Salvamar, which produces iron-ore pellets, published through the GCO.[130] The authors of the study note that in 1990 the company analysed 250 profitable corporations and selected 25 to emulate or model its own operations on. Samarco Salvamar developed projects and integrated management models into its practices, including environmental ones, by drawing from the experiences of the 25 companies that it had selected for study. This suggests that although Samarco Salvamar was rational in how it approached its operations, its decisions were constructed by what it observed in the 25 corporations that it studied.

The potential for ideas to diffuse amongst corporations which are participants in the GC is likely to be strong. This is because of the pressure they must feel not only to perform competitively to increase profits but also to act in more socially responsible ways, particularly in relation to the environment. The demonstration of successful innovations in relation to environmental initiatives which can generate income for corporations is likely to have an important causal effect on them. There are at least three arguments about how corporations as participants in the GC are likely to increase, because of the social influence that others could have directly or

---

[128] See the discussions associated with figure 7.4 summarising some of the findings of the 2007 McKinsey & Company report, particularly in relation to the pressures on CEOs to be socially responsible.

[129] Braithwaite and Drahos, above n 122, 30.

[130] Cláudio Bruzzi Boechat and Fundação Dom Cabral, 'One company, One Idea: Samarco Salvamar Project' in Global Compact Office, *Experiences in Management for Sustainability* (2003) United Nations Global Compact 37–52, www.unglobalcompact.org/docs/news_events/8.1/exp_man.pdf.

diffusely on the GC participants, the potential for the diffusion of ideas through emulation or modelling:[131]

1. The GCO in a 2007 survey reported that 63 per cent of the companies that had joined the GC did so for the general public to 'increase trust' in them.[132] Also, 53 per cent were there to network with others. This compares to 20 per cent which wanted to improve their market access for their businesses. This suggests that companies see the potential to socially influence others in their engagement with the GC. The risks of an adverse effect are also high. For instance, misrepresentations by companies might lead certain stakeholders like NGOs to label them as bluewashing their image, which might in turn impact on their image internationally.[133]
2. The GCO has reported that in 2007, 69 per cent of its corporate participants were actively communicating with it on the progress they were making towards implementing the core principles of the GC.[134] As previously discussed,[135] these reports are publicly available for comments from the GC's stakeholders and are published on the official web site of the GC for the general public also to view. Competitors of corporations that submit a COP can easily view their reports and assess how they perform in terms of implementing the environmental principles.
3. Some corporations themselves take an active role within the GC framework in diffusing ideas and innovations that are consistent with the core principles. For instance, the BMW Group has put itself forward to share its work through case studies on how its operations and those of its subsidiaries implement the GC principles.[136] A case study produced by Designworks, a subsidiary of BMW in the United

---

131 This is not to suggest that there are no other opportunities for diffusion of ideas and innovations within the GC framework. These are just three important ones that an analysis of the GC framework reveals as potentially useful for the purposes of diffusion.

132 Global Compact Office, above n 7, 11.

133 Bigge, above n 6.

134 Global Compact Office, above n 7, 51.

135 See above section 7.2.

136 The first case study was produced by the Center for Responsible Business at the University of California, Berkeley: Kellie McElhaney, Michael Toffel and Natalie Hill, 'Designing Sustainability at BMW Group: The Designworks/USA Experience' (Working Paper No 4, Center for Responsible Business, University of California, 2002). The second report is part of a case study series produced by the Global Compact Office, *From Principles to Practice*, above n 41.

> States, highlights its sustainability management system. It points out that it produced the study to 'share its practices with other companies interested in the Global Compact'.[137] Importantly, in another case study, a medium-sized company, Yawal System, in Poland, highlights how showing leadership had potential to contribute to diffusion. Yawal System sees the dissemination of 'information and experiences gained with other local companies' as acting in an environmentally responsible way.[138]

In these direct and diffuse interactions that corporations might have with others the role and function of principles as part of the social process of diffusion and the emulation or modelling of what some do become more apparent. Ideas or innovations are more socially influential and will potentially diffuse across the GC framework. This is because they are meaningful for actors either because they view them through the cognitive frame established by the abstract nature of the GC principles or because they fit within their understanding of the qualitative dimensions of the open-textured concepts in the GC principles. The social responsibility of corporations for the environment is bound to be a relative rather than an absolute idea, given their limited practices in this area discussed earlier in this chapter.[139] As a result they are more likely to engage with ideas and innovations that fit within a cognitive frame that establishes a level playing field for themselves and their peers. This does not suggest that the environmental principles of the GC harmonise or necessarily bring about a convergence of practices.[140] Actors who actively engage with the principles give meaning to them by adopting practices that are more consistent with their own corporate image and profile. The cognitive frame established by the principles is more likely to cause a range of ideas and innovations which might suit a greater diversity of corporations emulating them than if rules and standards, which are more defined, were used as part of the GC. In fact, it has been suggested that '[d]iffusion is far less deterministic and much more vulnerable to local conditions than the systemic view suggests'.[141] That is, the practices of corporations are less likely to imitate the

---

137 McElhaney, Toffel and Hill, above n 136, 14.

138 Global Compact Office, above n 101, 61.

139 See the discussion in section 7.3 and also figure 7.5.

140 Although this chapter discusses diffusion in the context of collective culture and learning it must be pointed out that this is not the same as suggesting that corporations, through the GC, are harmonising their approaches to environmental issues.

141 Eliason and Goldman, above n 118, 8.

exact ideas and innovations that others promote but yet remain within the frame established by environmental principles. This potentially engages a greater diversity of participants in a process which is no less likely to instantiate a collective culture of the kind described above.

It is possible to illustrate the strength and significance of this interplay between environmental principles and diffusion in terms of how it promotes a diversity of social practices through two different examples of innovative practices of corporations. Both these examples fit within the frame vaguely established by Principle 8 relating to corporations taking responsibility for the environment. These two examples have been distributed through one of the structured learning environments that the GCO has created.

The first is in a case study distributed by the GCO relating to British Petroleum (BP), which in 1998 committed itself to reducing Greenhouse Gas Emissions (GGE) to 10 per cent below 1990 levels by 2010.[142] Each of the Business Units (BU) of BP are given annual emission allocations. In 2000, one of its BU upgraded its activities which would cause it to exceed its allocated GGEs. It bought allowances from another BU of BP that had recently upgraded a furnace which had given it some spare allowances.[143] In this example, the potential for growth did not push BP to adopt an alternative preference of increasing its GGE.

In a different case study, also distributed by the GCO, Eskom, the largest supplier of electricity in Africa, is noted in 2000 to have introduced a demand side management (DSM) system giving customers the option to use less electricity during peak periods and to save money as a result.[144] With drops in consumption this would significantly delay the need to upgrade its networks and build additional power plants.[145] Eskom established a DSM department to educate the community about the potential to consume electricity sustainably. In reducing electricity consumption Eskom also reduced the total GGEs from power stations.[146] The environmental gains from this initiative were significant, given that the electricity

---

142 Global Compact Office, above n 101, 62.

143 Ibid.

144 Mokhethi Moshoeshoe and D Sonnenberg, 'Eskom and the Flexible Hot Water Load Management Control System' in Global Compact Office, *Experiences in Management for Sustainability* (2003) United Nations Global Compact 65–74, www.unglobalcompact.org/docs/news_events/8.1/exp_man.pdf.

145 Ibid,71.

146 Cornis Van Der Lugt, 'Introduction' in Global Compact Office, *Experiences in Management for Sustainability* (2003) United Nations Global Compact 9, www.unglobalcompact.org/docs/news_events/8.1/exp_man.pdf.

sector in 2003 contributed to 15 per cent of South Africa's gross domestic product.[147] Its corporate identity would have probably matured into a consumer oriented transnational corporation. Other corporations may not only model themselves on Eskom when faced with problems that require financial solutions but may proactively consider developing practices which would enhance a similar market presence to Eskom.

Comparing these two examples highlights the different approaches taken to reducing GGEs. Although their reasons for doing so were different, both companies are able to claim greater relative responsibility for protecting the environment. BP made internal changes to how it managed its business units, whereas Eskom sought to change consumer demand for its products. This suggests that both activities would be meaningful to different corporations because Principle 8 is able to accommodate a variety of ways that ideas could diffuse across to other corporations who, like them, are seeking to modify their practices.

Some might argue that the cognitive frame is too broadly established through the environmental principles because they encourage diffusion that is only weakly connected to the kind of stewardship that matters for protecting the environment and depletion of natural resources. That is, corporations may use the breadth of the principles to legitimise emulating practices that are less costly to them. In the context of collective learning of a new culture described above, this is not necessarily an issue for a longer term view of the kind of role or function that environmental principles play within the GC. This argument however does not deal with the potential that the specific properties of innovations have to cause diffusion of certain ideas.[148] That is, the presence of certain characteristics in ideas and innovations predisposes them to being more easily diffused than others. In this way environmental principles are able to frame certain types and kinds of ideas that more actors want to emulate or model themselves on. The example discussed in the next section relates to this point because it shows how the characteristics of an idea can socially influence others, not just because they are cognitively framed by the environmental principles of the GC.

---

[147] Moshoeshoe and Sonnenberg, above n 144, 72.

[148] See Jorgens, above n 119, who at 254–255 has argued that the following are important for diffusion:

> (1) the existence of international or transnational channels of communication through which information on policies in other political constituencies can be communicated; (2) the specific properties of policy innovations; (3) the specific structure of the problem that a given policy innovation is expected to deal with; and (4) the national capacities for adopting particular policy innovations.

7.4.1 Separating Innovation from the Local Context

It has been argued that diffusion is most successful when policy innovations can be separated from the 'specific national contexts in which they evolved'.[149] This is likely to be even more important in the case of the GC, which engages corporations from 116 countries around the world.[150] In such cases the ideas in themselves can socially influence corporations whether or not they are financially beneficial to them in the short term. For instance, this can be seen through a number of case studies compiled for the purposes of distributing them through the GCO, where there are strong references to transnational and other kinds of corporations working with external stakeholders like consumers, non-governmental organisations and local communities. For instance, Skanska, a Swedish construction company, when operating in Sri Lanka realised that its success in achieving compliance with ISO 14001 meant that it had to run 'an awareness-training program for all employees from the beginning of the project'.[151] However, in the same case study Skanska is criticised by comments through interviews suggesting that 'Skanska could have ensured greater credit for its environmental work had it engaged more with local communities and NGOs in dialogue'.[152] It cannot be assumed that corporations would necessarily pay attention to the details of this kind of case study published through the GC web page. However those interested in implementing ISO 14001 will find this criticism of Skanska to be useful for their own management practices.

In an alternative case study, the efforts of Shell International Ltd are highlighted by the GCO in the way it consulted with local and international non-governmental organisations and indigenous peoples during

---

149 Ibid, 256. See also Paul DiMaggio and Walter Power, 'The Iron Cage Revisited: Institutional Isomorphism and Collective Rationality in Organizational Fields' (1983) 48(2) *American Sociological Review* 147; David Strang and John Meyer, 'Institutional conditions for diffusion' (1993) 22 *Theory and Society* 487.

150 Global Compact Office, above n 7, 8.

151 Hans De Geer and Daniel Stridsman, 'In Search of Ecological Excellence: Skanska's Environmental Management at the Kukule Ganga Hydropower Project' in Global Compact Office, *Experiences in Management for Sustainability* (2003) United Nations Global Compact 13–36, 19, www.unglobalcompact.org/docs/news_events/8.1/exp_man.pdf. For information on ISO 14001 as an environmental management tool see the following web site sponsored by the International Organization for Standardization, *ISO 14000 Essentials*: www.iso.org/iso/iso_catalogue/management_standards/iso_9000_iso_14000/iso_14000_essentials.htm.

152 Van Der Lugt, above n 146, 7.

an appraisal and exploration operation in the Camisea region in Peru.[153] Shell also used the Smithsonian Institute, a corporate/non-governmental organisation funded body, to conduct the biodiversity assessment required for the area. Shell International Ltd is quoted as seeing these measures as building trust between different groups to better deal with operational issues and to make 'positive contribution to conservation'.[154] This level of engagement with local communities is also clearly visible from other companies such as the ones in the oil and gas industry the impact of which on biodiversity loss is most significant.

Both the precautionary approach and the need to promote greater environmental responsibility as abstract concepts point to an episteme requiring that companies manage 'human beings in their interaction with the environment'.[155] In the case of the precautionary principle or approach, engaging with local communities and non-governmental organisations will allow accurate environmental impact assessments to be prepared before a project begins. It will also assist with monitoring and precautionary measures to be taken during a project. In the case of the principle relating to environmental responsibility the GCO has pointed out that corporations that have decided to adopt this principle must 'contribute to environmental awareness in company locations', and 'communicate with the local community'.[156] The GC relies on Agenda 21 to justify its adoption of the principle of environmental responsibility. Agenda 21 also confirms the view that businesses need to foster 'openness and dialogue with employees and the public'.[157] From these case studies it would not be difficult to see how corporations entering new jurisdictions could benefit from engaging with local communities; delegating responsibility for aspects of the environmental assessment work to them; and collaborating with local NGOs to ensure that appropriate precautionary measures are taken based on local know-how. Both case studies discussed above would serve as useful models for companies to consider while ensuring they engage with the environmental principles of the GC.

This discussion highlights the potential of the environmental principles, cognitively to frame the advantages that come with corporations emulating or modelling themselves on what others do. The potential breadth of ideas and innovations that corporations can adopt while they remain

---

153 Global Compact Office, above n 101, 56.

154 Ibid.

155 Van Der Lugt, above n 146, 6.

156 Global Compact Office, above n 101, 59.

157 A copy of Agenda 21 is available at www.un.org/esa/sustdev/documents/agenda21.

engaged with the principles is clearly an issue. However this notion is tempered by the idea that diffusion is in itself a much more complex process which can lead corporations to be socially influenced to adopt practices that requires more from them than if they were left to their own devices.

7.5 IDENTITY AND SOCIAL INFLUENCE IN THE CONTEXT OF THE GLOBAL COMPACT

In addition to diffusion it is also arguable that through its mandatory requirements for reporting the GCO is also helping to define the type of social identity that some corporations might be willing to adopt as participants of the GC. Wendt in his *Social Theory of International Politics* describes the potential for types of identities to develop internationally.[158] For instance, an agent could be typified as a corporation because of the behavioural traits that it might share with other entities like it. A type of identity exists because actors share certain characteristics with other actors which have a social meaning in particular contexts. The social meaning of a particular type of identity may develop in a variety of ways including, for instance, 'less formal membership rules that define what counts as a type identity and orients the behaviour of Others toward it'.[159] To identify something as a particular type of agent does not mean that it is 'that' in all circumstances. If a type of identity is defined culturally then all agents, including corporations, can identify with different cultural forms in varying ways and at different times.[160] It is the cultural form which an agent takes up as its identity which also shapes its interests and preferences.[161]

The idea that the GC seeks to create 'corporate citizenship' through the UN is consistent with generating a type of a corporation that is a participant in the developing framework.[162] The GCO in a 2006 study highlighted that 'the percentage of companies communicating

---

[158] Alexander Wendt, *Social Theory of International Politics* (1999) 225. Wendt's book is primarily concerned with states, but the theories of symbolic interactionist he draws on originate from a variety of different sources in the literature on social psychology and arguably have equal relevance to other actors like transnational and other kinds of corporations.

[159] Ibid, 226.

[160] Ibid, 230.

[161] Ibid, 231.

[162] On the use of the idea of 'corporate citizenship' see the above discussion in section 7.2.

progress increases with the length of their participation in the Global Compact'.[163] In this report, the GCO produces figures that show that 95 per cent of corporations have submitted a COP when they have been participants in the GC for six years.[164] This is comparable to the 27 per cent who submit a COP after having been members of the GC for two years.[165] These figures suggest that membership of the GC is taken more seriously the longer the corporation remains a participant in the framework. It also indicates that citizenship or the type of identity for corporations seems to matter, given the higher percentage of established members completing a COP. From this, it is arguable that the COP creates a citizenship test for corporations which wish to remain active participants in the GC.

As discussed earlier in this chapter, a COP must state what a company has done in support of the GC. [166] It must describe practical actions it took in the reporting year to implement its core principles. It must also measure outcomes or specify what it expects them to be. Finally, the chief executive, chairman or some other senior executive of the company must cover the COP that is publicly distributed and put on the GC web site with a letter or a statement giving the GC its continuing support.[167] For instance, in the covering letter of the COP submitted by Bharat Heavy Electricals Limited (BHEL),[168] the largest engineering and manufacturing company in India, the Director for Human Resources wrote the following:

> BHEL is an environment [sic] friendly company in all its activities, products and services beside providing safe and health working environment to all its stakeholders. . . . The company reiterate [*sic*] its commitment to United Nation's Global Compact Programme and the set of core values enshrined in its ten principles on human rights, labour standards, environment and anti-corruption and intent [sic] to advance G.C. principles within the company's sphere of influence and has made it part of the strategy, culture and day-to-day operations.[169]

---

[163] Global Compact Office, *Leading the Way in Communication on Progress*, above n 44, 5.

[164] Ibid.

[165] Ibid.

[166] In relation to the COP see the earlier discussion in section 7.2.

[167] Global Compact Office, *Policy for 'Communication on Progress'*, above n 44, 2–3.

[168] On BHEL's COP uploaded on the official website of the GC see Global Compact Office, *Participants and Stakeholders,* United Nations, www.unglobalcompact.org/ParticipantsAndStakeholders/search_participant.html?submit_x=page.

[169] Bharat Heavy Electrical Limited, *BHEL's Commitment In Support of UN's Global Compact Programme* (2006) United Nations Global Compact,

It might be argued that the process of committing to the GC and having to reaffirm this annually seems like a performance. This process is important however in the symbolic effect it has in creating a particular type of corporation. Importantly, for companies that have set out their COP in this way it is apparent that their commitments and how they have implemented them are framed through the core principles.[170] What this discussion suggests is that the type of corporation which is created through the GC is managed by environmental principles creating for it the terms of its identity within the particular framework. This type of identity is important for instantiating a collective culture that values the future through shifts in attitudes to consumerism and production.

7.6 CONCLUSION: FUNCTION AND ROLE OF ENVIRONMENTAL PRINCIPLES, SOCIAL LEARNING AND THE GLOBAL COMPACT

From a survey of around 100 Fortune Global 500 companies Ruggie noted that 'companies generally do not "adopt" verbatim any of the human rights norms in voluntary codes of conduct developed internationally including the GC principles'. [171] Several indicated in their optional responses that while they were 'influenced by' or 'support' these instruments, their policies do 'not explicitly adhere' to or 'explicitly reference' them.[172] This study did not extend to the environmental principles of the instruments used for the survey, but it is highly likely that a similar response would be given for the environmental principles of the GC which, as discussed above, are abstract and open-textured. Given this possibility, an examination of the ability of environmental principles to interplay with processes that emphasise social influence can reveal more about their potential role and function within the GC framework. This is important, given the success of the initiative in establishing a strong membership base amongst

---

www.unglobalcompact.org/data/ungc_cops_resources/6D688856-BCEB-4600-AB86-3F27657C9EAB/COP.pdf.

[170] For example, in a study compiled by the GCO to highlight samples of what other corporations do, the style of COP used by Grupo Pão de Açúcar suggests that principles shape how its reporting is done and presumably the way its goals are also established for the following year: Global Compact Office, *Leading the Way in Communication on Progress*, above n 44, 18.

[171] John Ruggie, *Human Rights Policies and Management Practices of Fortune Global 500 Firms: Results of a Survey* (2006) 30, www.reports-and-materials.org/Ruggie-survey-Fortune-Global-500.pdf.

[172] Ibid, 5.

corporations and other stakeholders. It means that less focus needs to be placed on compliance which discounts the value of the long-term nature of collective learning and macro cultural change in international politics,[173] and regulation of the behaviour of actors resulting from it.

Importantly, it cannot be presumed that environmental principles have established the requisite intersubjectivity to determine corporate practices. This is so particularly because the international community has not successfully engaged corporations in shaping how issues of public concern are dealt with. Despite this, environmental principles of the GC framework may through social influence give corporations the opportunity to create meaning from them rather than intersubjectively share in their understanding of protecting the environment from particular patterns of consumerism or production. That is, as abstract or open-textured norms they may not establish the precise logic for what is the appropriate thing to do, but this is not to suggest that initiatives to shift corporate approaches in favour of taking precautionary measures, for instance, are not useful. Changes in the individual preferences of corporations of whatever kind are an instantiation of an emerging culture at the international level. That is, to shift a particular cultural understanding of consumerism and production corporations have to engage in collective learning experiences which can only begin by socialising them into changing their preferences in favour of taking initiatives towards environmental stewardship.

It is in the context of this approach to collective learning that this chapter has explored the role and function of environmental principles within the GC framework. In focusing on social practices of particular communities of actors the particular role and function of environmental principles are more concretely defined and examined. The characterisation of environmental principles as cognitively framing the diffusion of ideas and innovations and the type of ways that corporations could identify themselves in relation to others explains how they can collectively learn to take ecological approaches to consumerism and production.

Modelling and emulating what others do can happen because of the social influence that competition has on corporations which are judged in comparison to others like them. Also, certain ideas and innovations have greater potential to be diffused because of their particular qualities which can influence some corporations. This is because they can be abstracted from the local environment where they were learnt in the first place. Environmental principles can frame how corporations may react to social influences by making certain things appear rational or useful to them. For

---

173 See chapter 2, section 2.2.1 for a discussion of this idea.

instance, competition could be established on a more level playing field if corporations know that it is their interpretation of the three environmental principles of the GC that they would get judged on by the increasingly sensitive attitudes that the general public have to environmental stewardship. Alternatively, a particular idea could appear even more useful to a corporation that can see how the international community has come to contextualise it in terms of protecting the environment. Engaging local communities for instance in a project and in environmental impact assessment studies may seem even more rational or useful because it is consistent with taking a precautionary approach.

Further, the abstractness or the open-textured nature of the environmental principles means that corporations that do not identify exactly with how others view stewardship can still engage with and be part of the GC framework. The greater the number of participants which accept the citizenship test of submitting COPs the more likely it is that these corporations will also be instantiating the cultural change that has been discussed in this chapter. This is because of the way that the terms for the type of identity that the GC seeks to encourage emerge from the corporations interpreting the environmental principles for themselves. The public nature of the submissions of COPs exposes corporations to the social influence that others can have on them. The abbreviated and abstract presentation of the precautionary approach in Principle 7, for instance, means that stakeholders and the general public can potentially draw from a much wider range of knowledge areas to influence corporations. Similarly, the open-textured nature of the term 'responsibility' in Principle 8 means that the entire gamut of chapter 30 of Agenda 21 can be deployed in ways that will help socially to influence corporations through analysis of their COPs. This is not to suggest that the environmental principles cannot provide the basis for abuse, but by establishing the basis for their interplay with particular kinds of social influence the GC framework provides a valuable opportunity for corporations to take seriously the idea of instantiating a culture that will exert structural pressure on others that would never align themselves towards taking social responsibility.

8. Arguments, bargains and the disposal of carbon dioxide at sea

8.1 INTRODUCTION

In an empirically based work on the role of arguing and bargaining in multilateral negotiations, Ulbert, Risse and Müller noted that it was more likely for actors to argue and be open to persuasion if they were less certain about the nature of the problems being discussed and their own interests and preferences.[1] In a very different but significant study, Wendt argues that certain factors actively and efficiently contribute to actors collectively identifying with each other.[2] These factors he suggests are interdependence, common fate and homogeneity.[3] The similarity between these authors lies in the way they describe the social possibilities for actors to develop their preferences during their interactions with others. Scholars in international relations have been suggesting that studies of multilateral negotiations need to focus more on the dynamics of the social rather than the material context within which dialogue and communications take place amongst actors.[4] This is not just for ideological reasons but

---

1 Cornelia Ulbert, Thomas Risse and Harald Müller, 'Arguing and Bargaining in Multilateral Negotiations' (Paper presented at the Conference on Empirical Approaches to Deliberative Politics, France, 21–22 May 2004) 14. This idea is not new and was initially discussed in the context of social constructivist literature in Jeffrey Checkel, 'Why Comply? Social Learning and European Identity Change' (2001) 55(3) *International Organization* 553.

2 Alexander Wendt, *Social Theory of International Politics* (1999), 343–66.

3 Ibid.

4 The term material context is used for instance by Ulbert, Risse, and Müller, above n 1, 4, who refer to such things as the 'power resources of participants, their cost–benefit calculations and material factors that heighten the credibility of commitments or make side payments possible.' This idea is commonly shared amongst constructivists who argue for an ontology that focuses on social as well as material contexts in international relations. See, in particular, Thomas Risse, '"Let's Argue!": Communicative Action in World Politics' (2000) 54(1) *International Organization* 1; Harald Müller, 'Arguing, Bargaining and All That: Communicative Action, Rationalist Theory and the Logic of Appropriateness in International Relations' (2004) 10(3) *European Journal of International Relations*

rather because this focus helps to reveal more of what is going on at the international level. A focus on the social context also places emphasis, for instance, on the relevance of intersubjective structure which can include norms as compared to the power resources of key players.[5] This means that the nature of dialogue and its interplay with norms are important for change and social learning amongst groups because they are dimensions of the social context at the international level. This is not just from the technical perspective of solving problems but also in terms of creating common and collective preferences by interacting with and understanding others.

Within multilateral negotiations or conferences of parties to international agreements, actors could therefore be seen as bargaining or arguing, depending on whether they are acting strategically or communicatively.[6] Arguing refers to situations where actors who interact seek to understand others and the situation they are in rather than being concerned only about how they can further their self-interest.[7] Deitelhoff and Müller go as far as to identify arguing as a constant rather than a variable of multilateral negotiations.[8] Lose supports this idea in a different way by writing that in the world of diplomacy the interactions of actors are by nature 'oriented towards achieving a collective understanding of the interests and goals that states wish to translate into norms and rules guiding interstate

---

395; Cornelia Ulbert and Thomas Risse, 'Deliberately Changing the Discourse: What Does Make Arguing Effective' (2005) 40 *Acta Politica* 351; Andrew Linklater, 'Dialogic Politics and the Civilising Process' (2005) 31(1) *Review of International Studies* 141; Leonard Schoppa, 'The Social Context in Coercive International Bargaining' (1999) 53(2) *International Organization* 307; Jeffrey Checkel, 'Social Construction and Integration' (1999) 6(4) *Journal of European Public Policy* 545; Checkel, above n 1; Friedrich Kratochwil, *Rules, Norms, and Decisions: On the Conditions of Practical and Legal Reasoning in International Relations and Domestic Affairs* (1991).

5 Ulbert and Risse, above n 4, 354.

6 This is a very simplified version of the idea which is discussed further in section 2 of this chapter. Ulbert and Risse, ibid at 352, suggest that arguing is a 'hard case' in the context of multilateral negotiations because states mostly seek to achieve goals and to pursue their self-interest in that process.

7 On communicative action in international relations see, for instance, Risse, above n 4; Harald Müller, 'International Relations as Communicative Action' in Karin Fierke and Knud Jørgensen (eds), *Constructing International Relations: The Next Generation* (2001) 160–178; Müller, above n 4; Ulbert and Risse, above n 4.

8 Nicole Deitelhoff and Harald Müller, 'Theoretical Paradise – Empirically Lost? Arguing with Habermas' (2005) 31 *Review of International Studies* 167.

behavior'.[9] This suggests that actors can learn as a group to identify with certain preferences and approaches to issues because they have to communicate through arguing rather than just bargain and act strategically.[10] In this sense, arguing and persuasion are the steering mechanisms for change and social learning which need to be examined in the context of their interplay with norms.[11]

This chapter discusses the function and role of environmental principles in the context of negotiations at the international level and the work done at the Consultative Meetings of the contracting parties to the 1972 *London Convention* ('LC')[12] and the 1996 *Protocol to the London Convention* ('LP').[13] By tracing the development of the LP and the role of environmental principles within the ocean dumping regime this chapter seeks to better understand social learning in the context of the recent responses by the conferences of parties to the LC and LP to the issue of the sequestration of CO_2 and its injection into and storage in sub-seabed geological formations. In particular it examines the extent to which the discursive practices of actors resembled arguing rather than bargaining because environmental

---

9 Lars Lose, 'Communicative Action and the World of Diplomacy' in Karin Fierke and Knud Jørgensen (eds), *Constructing International Relations: The Next Generation* (2001) 179, 189.

10 Risse for instance has written that:

> [a]rguing constitutes a learning mechanisms by which actors acquire new information, evaluate their interests in light of new empirical and moral knowledge, and – most importantly – can reflexively and collectively assess the validity claims of norms and standards or appropriate behaviour.

Thomas Risse, 'Global Governance and Communicative Action' in David Held and Mathias Koenig-Archibugi (eds), *Global Governance and Public Accountability* (2005) 164, 164.

11 For connections between arguing and persuasion in terms of steering change see Thomas Risse, 'Global Governance and Communicative Action' (2004) 39(2) *Government and Opposition* 288; Ulbert and Risse, above n 4, 359.

12 *Convention on the Prevention of Marine Pollution by Dumping of Wastes and Other Materials*, opened for signature 29 December 1972, 1046 UNTS 120 (entered into force 30 August 1975) ('LC'). The agreement is more commonly known as the *1972 London Convention.*

13 *Protocol to the Convention on the Prevention of Marine Pollution by Dumping of Wastes and Other Matter*, opened for signature 7 November 1996, (1997) 36 ILM 1 (entered into force 24 March 2006) ('LP'). The term 'ocean dumping regime' will be used to refer to the collective impact of the LC, LP, and the resolutions of the Consultative Committee to both agreements. The LP as of 31 March 2008 had 34 contracting parties compared to the 82 that have signed and ratified the LC. See International Maritime Organization, *London Convention 1972*, www.imo.org/home.asp?topic_id=1488

principles, particularly the precautionary principles, interplayed with and constituted the social context of negotiations and dialogue amongst states. This is highlighted by comparing the possible reaction of contracting parties to the LC to the same issues that they face as parties to the LP.

This chapter has six sections. Initially, section 8.2 discusses how scholars in international relations who are moderate constructivists have conceptualised arguing in contrast to bargaining as another way to interpret the nature of negotiations at the international level. This section will also consider how arguing as a persuasive social process impacts on social learning. Section 8.3 describes the ocean dumping regime and environmental principles developed as part of it.

Section 8.4 discusses the development of the approach taken as part of the ocean dumping regime to the injection of sequestered CO_2 into the ocean and the sub-seabed. Section 8.5 examines the function and role of the precautionary principle in particular, in the social learning amongst contracting parties relating to the disposal of sequestered CO_2 in the ocean or the sub-seabed. The open-textured nature of the precautionary principle gave creative force to as well as constrained how actors sought mutually to understand each other. This would not have been possible without the strong lifeworld or intersubjectivity that distinguished what happened within the conferences of parties after the LP.

8.2 ARGUING AND SOCIAL LEARNING IN MULTILATERAL NEGOTIATIONS

To suggest that negotiations amongst actors are characterised by arguing means that there is in the relevant forums, like the conferences of parties to multilateral agreements, an 'exchange of reasons by participants who are oriented to reaching consensus' and who 'remain open to changing their minds if faced with better reasons'.[14] It is this intentional use of 'good reason' by transnational actors to search cooperatively for a 'common ground' which separates arguing as a communicative act from bargaining as a strategic activity.[15] The logic behind arguing as a form of communicative action is that an actor seeks not to influence others but to

[14] Jennifer Mitzen, 'Reading Habermas in Anarchy: Multilateral Diplomacy and Global Public Spheres' (2005) 99(3) *American Political Science Review* 401, 401. The works of Jürgen Habermas referred to by Mitzen include: *The Theory of Communicative Action: Vol 2* (1984); *Between Facts and Norms* (1996).

[15] Lose, above n 9, 184. The term 'communicative act' or action is another way of referring to arguing or communicative rationality.

reach agreement about something.[16] In this process what is important is for actors to seek to mutually understand one another by testing the validity of a variety of social claims that actors would make in relation to, for instance, the 'truthfulness of speakers', the 'truth of empirical assertions', or the 'rightness of normative claims'.[17] Arguing, then, is an activity that relies on the intersubjective connections amongst actors. Their interactions create and reproduce the intersubjective structures of meaning present within a forum, institution or other transnational social environments.[18]

Although the phraseology of argumentative persuasion in contrast to bargaining is new to the social constructivist literature in international relations, scholars within the so-called English School in international relations have also considered how actors communicate in international negotiations and why they might seek mutually to understand one another in an attempt to find common ground or to resolve their differences.[19] The English School, through scholars like Hedley Bull, Adam Watson and others, has consistently argued that negotiators are 'biased towards understanding other points of view and other needs'.[20] According to the English School, diplomatic practices are internationally structured for the purposes of encouraging argument and 'behaviour oriented towards mutual understanding'.[21]

Arguing as a 'social process' focuses on the nature of the communication amongst actors beyond whether they are solving their functional and technical problems by exchanging resources or balancing the power that exists amongst them. Since actors are engaged in dialogue they are also involved in a process of instantiating or generating 'collectively constructed intersubjective structures'.[22] It is not that actors enter negotiations with the

---

16 Jürgen Habermas, *Theorie des Kommunikativen Handelns: Vol 1 & 2* (1981) 85–133, cited in Lose, above n 9, 189.

17 On the connections between validity claims and arguing see, for instance, Ulbert and Risse, above n 4, 353.

18 Thomas Risse, '"Let's Talk!" Insights from the German Debate on Communicative Behavior and International Relations' (Paper presented to the Annual Convention of the American Political Science Association, Washington DC, 27–31 August 1997).

19 On the English School see, in particular, Andrew Linklater and Hidemi Suganami, *The English School of International Relations: A Contemporary Reassessment* (2006); Barry Buzan, *From International to World Society? English School Theory and the Social Structure of Globalisation* (2004).

20 Lose, above n 9, 193. See also Hedley Bull, *The Anarchical Society: A Study of Order in World Politics* (3rd ed, 2002), especially 165–178.

21 Lose, above n 9.

22 I have borrowed this phrase from ibid.

aim of changing their views or interest but that the relational nature of their interaction with others predisposes them to having to understand better the social situations they are in. In this process of trying mutually to understand one another, actors also better understand themselves, their beliefs and the intersubjective structures within which they can make claims to something. Risse has noted that:

> [a]rguing constitutes a learning mechanism by which actors acquire new information, evaluate their interests in light of new empirical and moral knowledge, and – most importantly – can reflexively and collectively assess the validity claims of norms and standards of appropriate behaviour.[23]

When communicating, actors will negotiate with each other and, where there is doubt or uncertainty, seek to clarify the meaning of what they or others think. In other words, they seek to 'understand' one another or the truth in the particular social situation. This is in contrast to pursuing a particular strategic end presuming that certain others will causally respond in particular ways because, for instance, they are in a less powerful position, or less resourced, in comparison.

For social learning to be effective, argumentative practices must draw from intersubjective structures or reification-like norms. In other words, communicative action or argumentative practices require a lifeworld before they can be effective or productive in the situation in which participants are engaging with each other. A 'common lifeworld' is a concept which comprises the 'crucial background condition for argumentative behaviour supplying common experiences with the world and its history as well as a common system of values and norms to which actors can refer in their communications'.[24] The concept of a lifeworld helps explain why actors are capable of arguing or pursuing a mutual understanding of problems or issues in particular circumstances. That states engage in diplomacy depends on concepts such as sovereignty and the role of a diplomat in international affairs. The basic intersubjective structures that are instantiated through the everyday participation of states in international diplomacy give support to the most basic lifeworld that prevents total anarchy.[25] This

---

23 Risse, above n 11, 288–289.

24 Risse, above n 4, 14.

25 On this point see Lose, above n 9, 194. The idea that there is a lifeworld internationally is still debated, given the dominance of sovereign statehood as a conceptual frame. For discussion of this issue see Mitzen, above n 14; and Antje Wiener, 'The Dual Quality of Norms and Governance Beyond the State: Sociological and Normative Approaches to "Interaction"' (2007) 10(1) *Critical Review of International Social and Political Philosophy* 47, 60.

example, however, refers to what can be called a 'fundamental collectivity' which states use as the basis for other kinds of activities.[26]

The lifeworld of a community, an institution or other organised group is in itself the resource actors have at their disposal to conduct meaningful discussions and negotiations with others. It gives them a sense of what it is appropriate to say or to do within particular contexts and in this way sustains what actors learn in particular contexts. That is, the lifeworld supplies not only the intersubjective structures that define who someone is in a particular situation but also the logic for what it is socially appropriate and acceptable to do in that context.[27]

In a practical sense, the lifeworld can shape participation and therefore social learning in a number of ways. Ulbert and Wisotzki have argued that international humanitarian law provided the intersubjective structures that helped give meaning to general norms which were used to argue for the prohibition of anti-personnel landmines.[28] Ulbert and Risse also highlight how the principles and general negotiating framework used in drafting the 1985 *Vienna Convention for the Protection of the Ozone Layer*[29] and the 1987 *Montreal Protocol on Substances that Deplete the Ozone Layer*[30] served as the 'reference models' for the 1992 *United Nations Framework Convention on Climate Change*.[31] In both these instances, the lifeworld already established within those contexts facilitated 'meaningful communication' amongst actors in the same or alternative settings. The lifeworld in these instances provided the background against which actors could better understand each other and therefore made persuasive arguments more probable and effective.[32]

How lifeworlds develop and whether they are effective in shaping

---

26 Lose, above n 9, 195.

27 On this see, for instance, Ulbert and Risse, above n 4, 354. On the concept that actors are driven to act according to the logic of appropriateness see, in particular, James March and Johan Olsen, *Rediscovering Institutions: The Organizational Basis of Politics* (1989).

28 Cornelia Ulbert and Simone Wisotzki, '(Pre-)structuring Argumentative Processes: the Impact of Institutional Settings on Banning Worst Forms of Child Labour and Landmines' (Paper presented at the 4th Pan European International Relations Conference, University of Kent, Canterbury, 8–10 September 2001), cited in Ulbert and Risse, above n 4.

29 Opened for signature 22 March 1985, 1513 UNTS 324 (entered into force 22 September 1988).

30 Opened for signature 16 September 1987, 1522 UNTS 3 (entered into force 1 January 1989).

31 Opened for signature 9 May 1992, 1771 UNTS 107 (entered into force 21 March 1994). See Ulbert and Risse, above n 4, 355.

32 Ibid, 356.

actors' learning experiences is relative to the particular situation which is being examined. In the case of multilateral environmental agreements (MEAs), states have often called for conferences convened by member states to negotiate what can and should be collectively done in relation to the particular issues before them.[33] Several authors have, however, argued against the usefulness of international conferences.[34] Haas, for instance, has argued that not every kind of conference of parties to an MEA has been capable of changing the beliefs or the practices of states in relation to the particular issues before them.[35] The vast philosophical differences in the responses of actors to environmental issues can complicate how lifeworlds have developed. This is similar to the argument above about the difficulties of establishing a lifeworld in a complex international society without a constitutionalised system of governance. For conferences to have an impact Haas has emphasised the constructivist bias favouring 'iterated interactions' amongst actors through, for instance, 'repeated preparatory meetings' for an international conference.[36] This is similar to suggesting that the interaction of actors needs to follow a particular pattern to develop an effective lifeworld amongst its participants. Others argue that continued dialogue in itself is capable of generating, sustaining and confirming intersubjectivity which can constitute 'the basic understanding that facilitates cooperation around a collectively defined aim'.[37] Whether a lifeworld of less significance exists internationally depends on the particular context studied. Even within international institutions which are seen as generating 'dense interaction patterns' the development of a lifeworld will depend very much on the normative frameworks which structure the interactions of the actors within it.[38]

---

[33] In relation to the formalisation of this mechanism as an institutional response to the development of multilateral environmental agreements see Robin Churchill and Geir Ulfstein, 'Autonomous Institutional Arrangements in Multilateral Environmental Agreements: A Little-Noticed Phenomenon in International Law' (2000) 94(4) *The American Journal of International Law* 623.

[34] See, for instance, Giulio Gallarotti, 'The Limits of International Organization: Systematic Failure in the Management of International Relations' (1991) 45(2) *International Organization* 183; Michael Barnett and Martha Finnemore, 'The Politics, Power, and Pathologies of International Organizations' (1999) 53(4) *International Organization* 699.

[35] For a consideration of the impact of international conferences on the environment see, for instance, Peter Haas, 'UN Conferences and Constructivist Governance of the Environment' (2002) 8 *Global Governance* 73, 76–77.

[36] Ibid, 76.

[37] Lose, above n 9, 195.

[38] Risse, above n 4, 15.

8.2.1 Identifying Arguments and the Role of Norms

It has been suggested that both bargaining and arguing express ideal types of interactions and the reasons why actors might say or do something cannot always be clearly established.[39] Arguing as an expression of communicative rationality is 'richer and more restrictive' in 'structure than strategic action' which is expressed in bargaining.[40] This difference is difficult to observe because, as stated for instance by Bernstein, the theoretical approach taken in terms of communicative action is 'reconstructive', meaning that one will not be able to hypothesise in relation to it and deduce evidence to prove it.[41]

Drawing on empirical research in international relations, some scholars have pointed out that communicative and instrumental reasoning are both present within particular contexts.[42] Even those who support the views of the English School relating to diplomatic negotiations and the mutual understanding that goes with them are quick to point out that this is not to say that 'diplomats empathize 100 percent and ignore the national interest in pursuit of international consensus'.[43] Risse, Müller and Ulbert have suggested that bargaining and arguing can be distinguished in analytical terms by proposing that the differences can be seen in 'modal', 'structural' and 'procedural' terms.[44] Referring to the structural differences between arguing and bargaining, Risse has noted that:

> arguing can be distinguished from bargaining through its triadic nature. . . . Bargaining actors assess the moves in negotiations based solely on their own utility functions including private information, while validity claims such as the truthfulness of speakers, the truth of empirical assertions, or the rightness of normative claims, recede in the background and are irrelevant for the bargaining situation. This is the dyadic nature of bargaining. Third parties might be present in bargaining situations as mediators or as enforcers of commitments.

---

39 Ronald Krebs and Patrick Jackson, 'Twisting Tongues and Twisting Arms: The Power of Political Rhetoric' (2007) 13(1) *European Journal of International Relations* 35.

40 Anita Kihlström and Joakim Israel, 'Communicative or Strategic Action – an Examination of Fundamental Issues in the Theory of Communicative Action' (2002) 11(3) *International Journal of Social Welfare* 210, 211.

41 Ibid, referring to Richard J Bernstein, *Beyond Objectivism and Relativism: Science, Hermeneutics and Praxis* (1983).

42 See, for instance, Deitelhoff and Müller, above n 8; Ulbert, Risse and Müller, above n 1.

43 Lose, above n 9, 193–194.

44 See Ulbert and Risse, above n 4, 352; Ulbert, Risse and Müller, above n 1; Risse, above n 11.

> But pure bargaining does not involve references to common values or to shared knowledge. Bargainers have to assess whether their counterparts can deliver on their promises or can issue credible threats, but the normative validity of the bargaining moves are irrelevant in this context.
>
> In contrast, arguing always involves references to a mutually accepted external authority to validate empirical or normative assertions. In international negotiations, such sources of authority . . . can be previously negotiated and agreed-upon treaties, universally held norms, scientific evidence, and other forms of consensual knowledge. In each case, there are different sources of external authority to which speakers can refer in order to validate their claims.[45]

Ulbert and Risse highlight the fact that in triadic forms of dialogue, '[s]peakers always refer to some external authority to make validity claims. This may be a set of norms, but it may also be a physical audience who is listening.'[46] However, their arguments about the use of norms as part of the triadic nature of dialogue can be better nuanced by acknowledging that some are abstract and open-textured in nature and can encourage participation from actors.[47] That is, the term 'logic of appropriateness',[48] which is used synonymously in international relations with 'doing the right thing' or 'rule-guided behaviour', presumes a high degree of certainty or lack of ambiguity in the structure and form of the norm itself. This means that when diplomats interact they seek to argue drawing on norms that have been collectively developed and interpreted. However, the dynamic of argumentative rationality in these instances is unlikely to enhance social learning because of the narrowness and bounded nature of more precise norms in the form of rules. That is, actors do not have the opportunity to interact for the purposes of creating meaning for themselves either in the particular context in which the rule is being applied or in the collective processes that guide the interpretation of the norm for the future. To use the language developed within international relations: although actors have a frame of reference for communicating with each other in triadic terms there is little room for rules to encourage reflexivity, which means that they are expected to follow a particular sequence of dialogue and reason-giving. Further, rules preclude the possibility that actors will be able to adjust the frames of reference which direct the course of communication between them.[49]

---

45 Risse, above n 11, 297–298.

46 Ulbert and Risse, above n 4, 357.

47 See Wiener, above n 25. See chapter 3 of this book in particular for arguments relating to how environmental principles are like this.

48 On this see, in particular, March and Olsen, above n 27.

49 See Ulbert and Risse, above n 4, 352.

As discussed in chapter 3 of this work, environmental principles are commonly constructed as abstract and open-textured, which in turn has the potential to create a degree of uncertainty in terms of what they mean in particular contexts. It is argued that, far from precluding or confusing argumentative discourse, their abstract nature has the potential to influence the dynamics of negotiation whereby actors are more likely to seek mutually to understand one another by generating new ideas or ways of socialisation with each other in the future. The abstract or open-textured nature of environmental principles has potentially varying effects on argumentative practices in the triadic structural sense described above. It can constrain as well as stimulate creative practices and engagements amongst the actors.

In one sense, writers have explored how uncertainty forces actors and creates the opportunities for them to 'coordinate their actions through communication oriented towards mutual understanding'.[50] Checkel has hypothesised that novel and uncertain environments are more likely to 'cognitively motivate' an actor to 'analyze new information'.[51] He identified the newness of an issue, a crisis and serious policy failure as factors that would lead to uncertainty and motivate actors to be persuaded.[52] Ulbert and Risse have also hypothesised that when actors were uncertain of their preferences or lacked knowledge there was greater room for them to be persuaded of something.[53] In these types of uncertain situations environmental principles as an episteme, abbreviated in an abstract form, can function as a necessary part of the triadic form of the dialogue amongst actors engaging in multilateral negotiations. Social learning emerges from the way that actors as a group develop interlocking beliefs about what to do next. Alternatively, environmental principles themselves can also create uncertainty for actors in terms of what they mean in particular situations. Within conferences held for parties to multilateral agreements, actors could see it as their social role to give meaning and shape to the vagueness and uncertainties that come with what environmental principles

---

50 This phrase is used by Lars Lose, above n 9, 189, but in the context of the anarchic system of states more generally rather than in relation to environmental principles more specifically.

51 Checkel, above n 1, 562.

52 Ibid. For an alternative view on how this hypothesis might also be crafted by rational choice theorists see Andrew Moravcsik, 'Bringing Constructivist Integration Theory Out Of The Clouds: Has It Landed Yet?' (2001) 2(2) *European Union Politics* 226. See Checkel, above n 1, for a counter argument to that of Andrew Moravcsik.

53 Ulbert and Risse, above n 4. See also Risse, above n 4, 19.

mean and how they should apply them in particular contexts. In this sense the uncertainty within the norms stimulates arguing as part of the triadic nature of dialogue within the conferences.

Alternatively, environmental principles as abstract and open-textured norms may, when employed in the triadic sense, also constrain actors in their interactions with each other. It has been argued that when actors negotiate in multilateral diplomatic environments some kind of a constraint is necessary to keep them 'committed to resolving their disagreements discursively'.[54] That is, there is a need to 'contain the instability of communicative action' or arguing within negotiations.[55] The presence of a strong lifeworld is one such stabilising force in international negotiations. Additionally, the ideas or discourses that environmental principles frame can also contain the instability that comes with using them. That is, although there is a great deal of uncertainty in relation to both the terms sustainable and development, they also manage to contain the kinds of arguments that actors will value in a particular context. The lifeworld supporting the use of the terms sustainable and development is extensive, but actors who draw on it will nevertheless be constrained by it because of the regularity of its use in the international community of states and the intersubjectivity or lifeworld within the situations in which it is being used.

The discussion in this section has examined how within multilateral negotiating frameworks actors argue instead of bargain. It describes and develops the idea that it is possible to differentiate arguing from bargaining on the basis of structural conditions present in the nature of negotiations itself. That is, along with the presence of a lifeworld, actors who engage triadically are more likely to argue than bargain. The important issue is the nature of the interplay between norms and the social processes of arguing, as opposed to whether it happens or not.

A variety of scholarly writings in international relations seek to explain how the preferences of actors can converge to create a common and shared understanding of issues. For instance, a process-oriented view would argue that, with the passage of time, actors within institutions are likely to develop a stronger sense of what they want as a group.[56] That is, changes in preferences are more likely if actors can interact over a longer period of time.[57] However, as Checkel has noted, these theories

---

[54] Mitzen, above n 14, 401.

[55] Ibid.

[56] Robyn Dawes, Alphons Van de Kragt and John Orbell, 'Not Me or Thee but We: The Importance of Group Identity in Eliciting Cooperation in Dilemma Situations: Experimental Manipulations' (1988) 68 *Acta Psychologica* 83, 83–97.

[57] See Jeffrey Checkel, '"Going Native" In Europe?: Theorizing Social

are underspecified and say little of the 'quality of the contact' that actors must have with each other for a sense of 'we-ness' to develop.[58] Risse has highlighted how communicative action, or arguing as the term was used in the discussion above, is 'crucial for social learning' because it is 'an important mechanism for socializing actors into new (international) norms and rules'.[59] This means that arguing as a social process specifies what according to Checkel is missing from those descriptions of how interlocking beliefs emerge within groups. This discussion has assessed how, instead of simply socialising actors into accepting norms and rules or encouraging their preference to converge, arguing, set in the context of an established lifeworld, can interplay with abstract and open-textured norms like environmental principles to create new ideas that belong to the group or can constrain how they respond to issues that are before them. In this sense what actors socially learn is how to do something as a group. The next section discusses these ideas in the context of the ocean dumping regime initially established by the LC in 1972.

8.3 DISPOSAL OF WASTE AT SEA

The disposal of waste at sea has long been an attractive option for states. The act of disposing wastes or any other matter into the ocean is generally referred to as dumping.[60] Ocean dumping is conceptually differentiated globally from direct land-based and vessel-sourced marine pollution. It is estimated that approximately 10 per cent of the pollutants which enter the marine environment are the result of deliberate dumping activities undertaken by coastal states.[61] Unlike some other forms of marine pollution, ocean dumping is an intentional or deliberate act which is therefore amenable to a considerable degree of administrative, legal, and technological control.[62]

Until the early 1970s, the international regulation of waste disposal at sea was limited to radioactive wastes under the 1958 *Geneva Convention on*

Interaction in European Institutions' (2003) 36 *Comparative Political Studies* 209, 210.

58 Ibid.

59 Risse, above n 11, 300.

60 Office for the London Dumping Convention at the International Maritime Organization, *The London Convention: The First Decade and Beyond* (1991) 44.

61 Ibid. See also John Kindt, *Marine Pollution and the Law of the Sea* (1986) 1087.

62 Ibid, 1087.

the High Seas.[63] The concern over the disposal of wastes at sea was not pursued any further until the 1972 Stockholm Conference when, as part of the Stockholm Action Plan, countries were asked to consider the need for 'an overall instrument for the control of ocean dumping'.[64] Since 1972 a number of multilateral agreements dealing directly or indirectly with dumping at sea have come into force. These include, for instance, the LC which was concluded in the same year as the Stockholm Conference, the 1982 *United Nations Convention on the Law of the Sea*,[65] and the 1986 *Protocol for the Prevention of Pollution of the South Pacific Region by Dumping*.[66]

The LC has widespread support internationally, with 84 contracting parties which make up 67.71 per cent of the world tonnage.[67] It is what Birnie and Boyle have called the most successful regulatory treaty of the 1970s.[68] The objectives of the LC are to 'promote the effective control of all sources of pollution to the marine environment' and in particular to prevent dumping of 'wastes and other matter'.[69] To this end its contracting parties must seek to harmonise their policies.[70] The general focus of the LC is on regulating the disposal of wastes and other substances in the

---

63 *Convention on the High Seas*, opened for signature 29 April 1958 (1958) 450 UNTS 82, (entered into force 30 September 1962) Art 25. See Daniel P Finn, 'Ocean Disposal of Radioactive Wastes: The Obligation of International Cooperation to Protect the Marine Environment' (1981) 21 *Virginia Journal of International Law* 621; Myres McDougal and William Burke, *The Public Order of the Oceans: A Contemporary International Law of the Sea* (1961) 864–867.

64 *Declaration of the United Nations Conference on the Human Environment*, UN Doc A/CONF.48/14/Rev.1 (1972). See also *United Nations Conference on the Human Environment: Draft Articles on Ocean Dumping*, UN Doc A/Conf.48/IWGMP II5 (prov) 15–21; (1979) 11 ILM 19.

65 *United Nations Convention on the Law of the Sea*, opened for signature 10 December 1982, 1833 UNTS 397 (entered into force 16 November 1994) ('*UNCLOS*'). It has been suggested that Arts 210 and 216 of *UNCLOS* codify customary international law in relation to the disposal of waste at sea: see Patricia Birnie and Alan Boyle, *International Law and the Environment* (1992) 320.

66 *Protocol for the Prevention of Pollution of the South Pacific Region by Dumping*, opened for signature 25 November 1986 (1987) 1982 UNTS 4 (entered into force 22 August 1990) ('*SPREP Dumping Protocol*').

67 See the official website of the International Maritime Organization, www.imo.org. World tonnage is significant because it represents the number of countries and the percentage of vessels that can potentially engage in disposing of waste at sea.

68 Birnie and Boyle, above n 65, 330.

69 *LC*, opened for signature 29 December 1972, 1046 UNTS 120 (entered into force 30 August 1975) Art 1.

70 Ibid, Art 2.

belief that the oceans have the capacity to assimilate waste. Annex I to the LC lists wastes and other matter which states must not dump at sea.[71] On the other hand, Annex II to the LC lists wastes and other matter that can be dumped at sea but which require a prior special permit from the state which is allowing the activity to take place in its waters.[72] However, contracting parties to the LC are allowed to dump anything they like that is not listed in Annexes I and II and have only to issue a general permit for doing so.[73]

The Consultative Meetings of the contracting parties to the LC have been successful in developing consensus around difficult issues relating to dumping.[74] The Government of the Netherlands in a document prepared in 2002 for the Consultative Meeting of the LC, has commented how the agreement has proven 'to be very much a living Convention, with the discussions at various meetings resulting in substantial changes to the practical implementation of its provisions'.[75] A summary of the progress made through the Consultative Meetings by contracting parties in protecting the marine environment from dumping was published for the first time in 1989.[76] Since 1975, contracting parties have met annually and reports of their Consultative Meetings are published through the International Maritime Organisation (IMO) which acts as the secretariat for the LC and now the LP.[77] Contracting parties to the LC, through the Consultative Meetings, have made amendments to the Convention, provided guidelines for implementing its provisions, and established reporting and compliance measures. These functions are particularly important because of the way these meetings seek to develop a shared understanding of what the ocean

---

71 Ibid, Art 4(1)(a), annex 1.

72 Ibid, Art 4(1)(b), annex 2.

73 Ibid, Art 4(1)(c), annex 1.

74 D Tromp, 'Foreword' in Office for the London Dumping Convention at the International Maritime Organisation, *The London Convention: The First Decade and Beyond* (1991). See also Birnie and Boyle, above n 65, 321. The Consultative Meetings are an administrative and policy making body of the Contracting Parties to the LC. They consist of states that have ratified the LC.

75 Government of Netherlands, 'Future Objective for the London Convention 1972 and the 1996 Protocol thereto' in Twenty-Fourth Consultative Meeting of Contracting Parties to the Convention on the Prevention of Marine Pollution by Dumping of Wastes and Other Matter, LC 24/14, 9 August 2002, annex, 9 [2.4].

76 See Office for the London Dumping Convention at the International Maritime Organisation, above n 74.

77 These reports are publicly available through national offices that administer the LC on behalf of member states. For instance the Australian Maritime Safety Authority represents the Australian Government in the meetings of contracting parties to the LC.

dumping regime represents and how it should regulate activities that are potentially dumping. Birnie has argued that in the case of the LC many terms are 'subjective', the 'operation of which would require interpretation' by contracting parties.[78] She writes that in the case of the LC where no dispute settlement process exists, members use meetings to 'resolve any conflicts' arising from the interpretation of terms used in the multilateral agreement.[79] That is, actors learn what is appropriate and what they want in the context of arguing about what they and others understand concepts to mean.

In 1996, a special meeting of contracting parties to the LC considered and adopted the LP to the LC. Drafting of the LP began in July 1993 by the Amendment Group of the Consultative Meeting of the contracting parties to the LC, which met again in May 1994 and April 1995 to complete the draft. The draft versions of the LP were considered by the LC's Scientific Group and also the LC's Juristic and Linguistic Group. The Consultative Meeting of the LC in their sessions of November 1993 and June 1994 approved and reported on the progress being made by the LC Amendment Group, and in July 1996 approved drafts of the protocol and called for the convening of a special session of the Consultative Meeting later that year to consider and adopt the LP. The LP is the most significant instrument to date in terms of regulating the disposal and incineration of waste and other matter at sea. The LP, which entered into force on 26 March 2006, presently has 34 contracting parties (25.59 per cent of the world tonnage).[80] This is a small number compared to the LC itself which has 84 state contracting parties (67.71 per cent of the world tonnage).[81] This does not indicate the real status and effectiveness of the LP internationally. This is because contracting parties to the *UNCLOS*[82] are required by Article 210 of that agreement to adopt rules that will protect the oceans

---

78 Patricia Birnie, 'Twentieth-century marine conservation conventions adaptable to twenty-first century goals and principles?: Part II' (1997) 12(3) *International Journal of Marine and Coastal Law* 488, 489.

79 Ibid. This is not to suggest that contracting parties cannot resort to other formal and informal means available internationally for resolving any disputes between them, including the Tribunal for the Law of the Sea and the International Court of Justice: see, for instance, Natalie Klein, *Dispute Settlement in the UN Convention on the Law of the Sea* (2005).

80 International Maritime Organisation www.imo.org/.

81 Ibid.

82 As of 9 May 2008, 155 states had signed and ratified the agreement: see United Nations Division of Ocean Affairs and the Law of the Sea, *Chronological lists of ratifications of, accessions and successions to the Convention and the related Agreements as at 09 May 2008*, www.un.org/Depts/los/reference_files/

from dumping in a way that is no less effective than what is expected from them internationally. The Division for Ocean Affairs and the Law of the Sea of the United Nations Office of Legal Affairs has identified the LC as containing the global rules and standards binding on contracting parties to the *UNCLOS*.[83] There is no reason why a similar argument cannot be made for the LP, particularly as the number of states ratifying it increase over time.

The LP's main objectives require that contracting parties 'prevent, reduce and where practicable eliminate pollution caused' not only by dumping but by incineration at 'sea of wastes and other matter'.[84] The LP prohibits the disposal of waste and instead lists a number of 'wastes or other matter' that parties can consider dumping at sea.[85] In developing the LP, contracting parties to the LC which participated in the Consultative Meetings introduced the idea of 'reverse listing' by prohibiting the intentional disposal of all waste and other matter at sea, with the exception of those that it lists in its Annex I.[86] The idea of only listing what is acceptable for states to dump into the ocean began as a conversation amongst contracting parties back in 1994. Initially, the idea of 'reverse listing' was argued as taking a precautionary approach to give the public confidence in the effectiveness of the ocean disposal regime.[87] The Japanese government

---

chronological_lists_of_ratifications.htm#The%20United%20Nations%20Convention%20on%20the%20Law%20of%20the%20Sea

[83] Report of the Second Meeting of the Amendment Group, Comments by the Division of Ocean Affairs and the Law of the Sea, United Nations Office of Legal Affairs, Amendment Group 1994 (2nd Meeting) LC/AM.2/3/1, annex I [8]. See also Alan Boyle, 'Marine Pollution Under the Law of the Sea Convention' (1985) 79 *American Journal of International Law* 347; Robin Churchill and Alan Lowe, *The Law of the Sea* (1988) 272.

[84] LP, opened for signature 7 November 1996, (1997) 36 ILM 1 (entered into force 24 March 2006) Art 2.

[85] Ibid, Art 1, lists these and includes:

> (1) dredged material; sewage sludge; fish waste, or material resulting from industrial fish processing operations; vessels and platforms or other man-made structures at sea; inert, inorganic geological material; organic material of natural origin and bulky items primarily comprising iron, steel, concrete and similarly un-harmful materials for which the concern is physical impact and limited to those circumstances, where such wastes are generated at locations, such as small islands with isolated communities, having no practicable access to disposal options other than dumping.

[86] Ibid, Art 4.1.1.

[87] See Report of the Working Group on Reverse Listing, in Seventeenth Consultative Meeting of Contracting Parties to the Convention on the Prevention of Marine Pollution by Dumping of Wastes and Other Matter, LC 17/14, 3–7 October 1994, annex 3. The Government of the Netherlands had argued that the

had instead argued that the 'reverse listing' scheme would be 'a total restructuring of the Convention'.[88]

The ocean dumping regime initially established by the LC, and which now includes the LP, is 'generally regarded as one of the more successful international legal instruments, having led to an apparently significant reduction in the volume and type of waste being dumped at sea'.[89] Before the 1992 United Nations Conference on Environment and Development, the fourteenth Consultative Meeting to the LC highlighted some of its recent organisational achievements when it commented that:

> [i]mportant refinements of allowable practices have been forged during discussions and debate in the London Convention's formal Consultative Meetings, with intersessional guidance from subsidiary bodies. Those refinements include: an indefinite moratorium on low-level radioactive waste dumping at sea, dating back to 1983 that is the subject of ongoing risk-related assessments; agreement in 1988 to proceed towards the termination of ocean incineration of liquid noxious wastes by 1994 or earlier; and a decision in November 1990 to phase out industrial waste dumping at sea by 1995.[90]

The manner in which the LP was drafted illustrates the strength and the lifeworld present in the institutional structures established within the IMO to deal with ocean dumping issues. This has meant that environmental principles, even as abstract and open-textured norms, could play an important role in the overall ocean dumping regime. This is because

---

benefits of the 'reverse listing' scheme were important because it was 'a practical implementation of the precautionary approach'. This was because 'the application of the "reverse listing procedure" should facilitate the use of the Waste Assessment Framework. An extra advantage of the "reverse listing procedure" is that Annexes I and II of the present Convention are made obsolete': note by the Secretariat, 'Inventory of Amendment Proposals' in Amendment Group, First Meeting, LC/AM.1/3, 10 June 1993, agenda item 3, [41].

88 Submission by Japan, 'Comments on the Proposed Amendments to the London Convention', Sixteenth Consultative Meeting of Contracting Parties to the Convention on the Prevention of Marine Pollution by Dumping of Wastes and Other Matter, LC 16/4/6, 8–12 November 1993, agenda item 4.

89 Submission by the Netherlands, 'Future Work Programme Implementation and Review of the Long-Term Programme' in Twenty-Fourth Consultative Meeting of Contracting Parties to the Convention on the Prevention of Marine Pollution by Dumping of Wastes and Other Matter, LC 24/14, 9 August 2002, agenda item 14.1.

90 Fourteenth Consultative Meeting of Contracting Parties to the Convention on the Prevention of Marine Pollution by Dumping of Wastes and Other Matter, LC 14/16, 30 December 1991, annex 5, 2.

of the depth of intersubjectivity that exists amongst actors within the institutional structure that supports the LC and the LP.

8.3.1 Environmental Principles in the Ocean Dumping Regime

Environmental principles did not play a direct role within the ocean dumping regime until 1991 when the Consultative Meeting to the LC adopted a resolution to apply a version of the precautionary 'approach' to the framework established by the LC.[91] In 1991, the Consultative Meeting adopted a resolution for the first time which required contracting parties when implementing the LC to be 'guided' by a 'precautionary approach to environmental protection'.[92] The use of the word 'guide' does not adequately convey the difficulty that the contracting states were dealing with in adopting the principle. This is seen for instance in the text of Resolution LDC.44(14) (1991) itself, which made reference to the contracting parties becoming aware 'that the quantity, diversity and complexity of chemical compounds entering the environment' made it difficult to 'determine the overall threat to the environment'.[93] The resolution also raised concerns about the need to 'limit contamination of the marine environment by wastes and other matter'.[94]

Since Resolution LDC.44(14) (1991), sub-articles 3(1) and 3(2) of the LP to the LC require respectively that states must apply a 'precautionary approach to environmental protection' and should 'bear the cost of pollution'. Article 3(3) of the LP replicates Article 195 of *UNCLOS* and what is essentially Principle 14 of the *Rio Declaration* by requiring that states should also avoid transferring damage from one part of the environment to another or transforming a particular type of pollution into another.[95]

The construction of the precautionary approach as open-textured and

---

[91] Resolution LDC.44(14), 'The Application of a Precautionary Approach in Environmental Protection within the Framework of the London Convention' in Fourteenth Consultative Meeting of Contracting Parties to the Convention on the Prevention of Marine Pollution by Dumping of Wastes and Other Matter, LC 14/16, 30 December 1991, annex 2.

[92] Ibid.

[93] Ibid.

[94] Ibid.

[95] Principle 14 of the *Rio Declaration,* which is almost identical to Art 195 of *UNCLOS,* states that 'States should effectively cooperate to discourage or prevent the relocation and transfer to other States of any activities and substances that cause severe environmental degradation or are found to be harmful to human health'.

abstract is confirmed by the drafting history of the various instruments. These include for instance Resolution LDC.44(14) (1991) and Article 3(1) of the LP. The open-textured nature of the principle goes deeper than whether the contracting parties should 'apply' or only be 'guided' by the precautionary approach. In the discussions during the Consultative Meeting that approved LDC.44(14) (1991), which represented the first use of the principle within the ocean dumping regime, it was noted that the precautionary approach was 'a new policy based on the recognition of the failure of past approaches to environmental protection, e.g. the "assimilative capacity" approach which therefore required new measures to be adopted. . .'.[96] This statement recognises the need to move away from past practices but expresses uncertainty about what the future direction of precautionary policy measures should be.

Article 3(1) of the LP as well as Resolution LDC.44(14) (1991) specify that precaution must be taken when 'there is reason to believe' that dumping of wastes and other matter into the sea is 'likely to cause harm even when there is no conclusive evidence to prove a causal relation between inputs and their effects'.[97] Contracting parties rejected Sweden's proposals during negotiations to replace the words 'likely to cause harm' with 'may cause harm'. The arguments put forward by Sweden for this alternative were based on the premise that the words 'may cause harm' better balance the differences between prevention and precaution.[98] The final text, in contrast to the alternative put forward by Sweden, narrows the range of measures that states can take as a precaution to protect future generations, because it implies that a situation is already at risk of harm. That is, it is not as abstract as the wording of the precautionary principle adopted in Article 15 of the *Rio Declaration*.[99] This difference, however, does not appear to have played out in the future direction taken by the contracting parties, and in fact the intersubjective meaning of the precautionary principle developed out of practice appears to value the potential of environmental harm in much more abstract terms.

---

96 Fourteenth Consultative Meeting of Contracting Parties to the Convention on the Prevention of Marine Pollution by Dumping of Wastes and Other Matter, LC 14/16, 30 December 1991, annex 2 [4.4].

97 *LP,* opened for signature 7 November 1996, (1997) 36 ILM 1, (entered into force 24 March 2006) Art 3(1) and Resolution LDC.44(14), above n 91, [1].

98 Proposal by Sweden in 'List of Documents Issued in Connection with the First Meeting of the London Convention 1972 Amendment Group', London Convention 1972 Amendment Group (First Meeting), LC/AM 1/INF.2, 19–23 July 1993, 15.

99 *Rio Declaration,* UN Doc A/CONF.151/5/Rev.1 (1992).

The inclusion of the polluter pays principle in Article 3(2) raised difficulties for some states that were concerned with liability and compensation. However, Article 3(2) is only a declaration of intention to apply the polluter pays principle, which is moreover confined to costs involved with issuing permits and enforcement of permit conditions. Also, an interpretation of Article 3(2) by the delegation of France, and supported by the Solomon Islands, suggests that the polluter pays principle applies only to primary sources of pollution. This means, for example, that Article 3(2) applies to those activities causing contamination of sediments to be dredged, but not to dredging and disposal activities themselves.[100]

The rest of this chapter draws on the use of the precautionary approach and its interplay with the nature of the discursive practices of the contracting parties in the context of the developing lifeworld within the ocean dumping regime that includes the LC and LP. In particular it assesses its meaning and function in the context of the injection of sequestered CO_2 into geological formations in sub-seabed repositories.

8.4 INJECTING CARBON DIOXIDE INTO THE SUB-SEABED AS WASTE DISPOSAL

In the past several years concerns over the increased CO_2 output from inland activities, such as the use of fossil fuels, has highlighted the potential impact such activities have on increasing levels of acidity in the ocean and climate change more generally. Increased acidity in the oceans will have consequent effects on marine ecology, and also the 'marine sink of carbon dioxide'.[101] The dire consequences of high levels of CO_2 have

---

[100] Eighteenth Consultative Meeting of Contracting Parties to the Convention on the Prevention of Marine Pollution by Dumping of Wastes and Other Matter, LC 18/11/Rev 1, note 12, 10.

[101] The United Kingdom submitted a report to the LC's Consultative Meeting in 2004 where they note that:

> [a] decrease in ocean PH will affect the carbonate chemistry of the oceans which may affect phytoplankton species composition, principally by disrupting calcifying organisms. Other non-calcifying organisms may grow in its place and impact the structure and processes occurring in the whole ecosystem. Other organisms that fix calcium carbonate may also be at risk due to lowering PH. Amongst these are corals and shellfish. Species composition is intimately linked to the strength of the biological carbon pump and hence the natural long-term retention of carbon into the deep-oceans may also be influenced.

Submission by the United Kingdom, 'Mitigating the Environmental Impacts on the Oceans Climate Change: Carbon Capture and Sequestration in the Marine

created much interest in the possibility of storing it in geological formations in the sub-seabed. Geological formations in the sub-seabed, which can be used to store CO_2, include oil and gas reservoirs that are no longer being used, deep saline aquifers, and un-mineable coal seams.[102] The sequestration of CO_2 and its storage in sub-seabed geological formations are not without risk to the integrity of the marine environment, given the possibilities of leakage which would be particularly harmful if the stored CO_2 was contaminated.[103]

Environment' in Twenty-Sixth Consultative Meeting of Contracting Parties to the Convention on the Prevention of Marine Pollution by Dumping of Wastes and Other Matter, LC 26/6/1, 1 September 2004, agenda item 6, annex [1.3].

[102] Submission by the United Kingdom, 'Carbon Dioxide Sequestration in Geological Structures: Carbon Dioxide Capture and Storage – A Factual Description and Glossary for Carbon Dioxide Discussion under the London Convention and Protocol', Twenty-Eighth Meeting of the Scientific Group of the London Convention, LC/SG 28/3, 3 March 2005, agenda item 3, [3.9]. In 2004, the United Kingdom government highlighted in a report to the Twenty-Sixth Consultative Meeting of the Contracting Parties to the LC that 95 per cent of potential storage of CO_2 in Europe would be in 'deep saline aquifers' and only 5 per cent would find their way into depleted oil and gas fields: see Submission by the United Kingdom, above n 101, [1.5].

[103] See, Note by the Secretariat, 'Carbon Dioxide Sequestration in Sub-Seabed Geological Formations: Development of Specific Guidelines for Assessment of Carbon Dioxide Streams' in Twenty-Ninth Consultative Meeting of Contracting Parties to the Convention on the Prevention of Marine Pollution by Dumping of Wastes and Other Matter 1972 & Second Meeting of Contracting Parties to the 1996 Protocol to the Convention on the Prevention of Marine Pollution by Dumping of Wastes and other Matter 1972, LC 29/4, 5–9 November 2007, agenda item 4.1. In respect of impurities in CO_2 that is captured for storage, Greenpeace International makes the following comment in a submission to the 2006 Consultative Meeting of the LC and LP:

> Captured carbon dioxide streams contain impurities which would have practical impacts on carbon dioxide transport and storage systems, as well as potential health, safety and environmental impacts additional to that of the carbon dioxide itself. Because of the low pressure, the large presence of nitrogen from air and the large scale of primary fuel combustion units, huge flows of gases are generated, containing not only carbon dioxide, N2, O2, and H2O, but also air pollutants such SOx, NOx, particulates, HCI, HF, mercury, other metals, and other trace organic and inorganic contaminants. The types and concentrations of impurities in carbon dioxide stream captured post combustion will depend on the fuel type, combustion process and type of capture process employed.

Submission by Greenpeace International, 'Carbon Dioxide Sequestration in Sub-Seabed Geological Formations: Consideration of Proposals to Amend Annex I to the London Protocol' in First Meeting of Contracting Parties to the 1996 Protocol to the Convention on the Prevention of Marine Pollution by Dumping of Wastes and Other Matter 1972, LP 1/6/2, 8 September 2006, agenda item 6.1, [1.1].

The application of the LC and the LP to the problem of having to reduce the natural and significant amounts of CO_2 that are dissolved into the ocean is complex. This is because the regime was designed to eliminate pollution by dumping rather than to encourage it after CO_2 had been captured inland or offshore. It was only in the context of excess gas that contracting parties had previously considered storing waste in offshore wells.[104] The use of the seabed for disposal or storage of waste had also only been brought up in discussions relating to the disposal of high-level radioactive wastes that might not have had a preferred safer home.

It was in 1992 that the contracting parties to the LC were first made aware of the possibility of disposing of CO_2 in the deep ocean waters by injecting it into the water column itself.[105] At that stage, on advice from the Scientific Group of the LC, the contracting parties by-passed the issue because of 'numerous uncertainties and limitations of the scientific understanding' of the process of disposing of CO_2 in deep ocean waters.[106] The idea of injecting CO_2 into 'oceanic deepwater' was later considered and dismissed as an option for the ocean dumping regime.[107]

The ocean dumping regime had, by the late 1990s, helped states to see value in waste minimisation and not to prioritise the fiction that the oceans had an unlimited capacity to assimilate it. The precautionary approach, in the context of the LC and its LP, had also emerged more as a signal to take preventative measures rather than to encourage activities which could be useful but with potentially disruptive impacts on the ocean and its

104 Seventeenth Consultative Meeting of Contracting Parties to the Convention on the Prevention of Marine Pollution by Dumping of Wastes and Other Matter, LC 17/14, 3–7 October 1994, [3.25].

105 *Report of the Fifteenth Consultative Meeting*, Fifteenth Consultative Meeting of Contracting Parties to the Convention on the Prevention of Marine Pollution by Dumping of Wastes and Other Matter, LC 15/16, 9–13 November 1992, [6.22].

106 Ibid.

107 Note by the Secretariat, 'Consideration of the Report of the Scientific Group: Ocean Storage of Carbon Dioxide' in Twenty-First Consultative Meeting of Contracting Parties to the Convention on the Prevention of Marine Pollution by Dumping of Wastes and Other Matter, LC 21/INF.2, 25 June 1999, agenda item 5. On this see also Submission by the United Kingdom, above n 102, [2.7]. See also, for instance, Spain's explicit rejection of this possibility in 2006: Submission by Spain, 'Carbon Dioxide Sequestration in Sub-Seabed Geological Formations: Consideration of Proposals to Amend Annex 1 of the London Protocol' in First Meeting of Contracting Parties to the 1996 Protocol to the Convention on the Prevention of Marine Pollution by Dumping of Wastes and Other Matter 1972, LP 1/6/1, 30 October–3 November 2006, agenda item 6.

sub-seabed. As a result, proposals for CO_2 capture and storage that sought actively to use the sub-seabed as a site for storage challenged established social understandings relating to dumping of wastes at sea. Importantly, the proposal had the potential to test the established social learning which had accompanied the regime since the precautionary approach had been adopted through Resolution LDC 44(13) in 1999. This is because the advantages of storage are significant but the possibility of CO_2 leaking into the marine environment and atmosphere is also high.[108]

In 2006 the Consultative Meeting of the LP and the LC considered a proposal by Australia (co-sponsored by France, Norway and the United Kingdom) to amend the LP allowing for the disposal of CO_2 into sub-seabed repositories. The proposal, which later formed the basis of the changes in the regime, was simple and general, and amended Annex 1 of the LP to allow the disposal into sub-seabed geological formations of streams consisting overwhelmingly of CO_2.[109] Since then, contracting parties have sought to complement this amendment by developing guidelines for states to assess whether to issue permits for the disposal of CO_2 into sub-seabed geological formations.[110]

The discussion below assesses how approaches vary between the LP and LC as to the concept of storing or disposing of CO_2 as well as areas within the ocean where states can do this. The significant differences in approach between the LP and LC to these two issues are important because of the way that the precautionary principle engages actors in learning how to manage waste disposal activities.

---

[108] See OSPAR Commission for the Protection of the Marine Environment of the North-East Atlantic, 'Placement of Carbon Dioxide in Sub-Sea Geological Structures' in Note by the Secretariat, 'Continuation of a Review of the Benefits of Carbon Dioxide Sequestration in Sub-Seabed Geological Structures and of New Information Relevant to the Protection of the Marine Environment', Meeting of the SG Intercessional Technical Working Group on Carbon Dioxide Sequestration, 3–6 April 2006, LC/SG-Carbon Dioxide.1/INF.2, Annex at 6; also available at www.ospar.org/eng/html/welcome.html.

[109] Submission by Australia (co-sponsored by France, Norway and the United Kingdom), 'Carbon Dioxide Sequestration in Sub-Seabed Formations: Consideration of Proposals to Amend Annex 1 to the London Protocol' in First Meeting of Contracting Parties to the 1996 Protocol to the Convention on the Prevention of Marine Pollution by Dumping of Wastes and Other Matter 1972, LP 1/6, 28 April 2006, agenda item 6.2.

[110] See, for instance, Note by the Secretariat, above n 103.

8.4.1 Carbon Dioxide Disposal, Storage, and Reverse Listing in Annex 1 to the LP

In 2006, Annex 1 to the LP was amended to allow for certain circumstances where CO_2 could be injected into sub-seabed geological formations.[111] This amendment regulates only certain kinds of CO_2 capture and storage because of the jurisdictional reach of the LC and the LP. For instance the disposal of waste generated offshore during normal operations on platforms or other man-made structures is not dumping according to the LP.[112] Also, injecting CO_2 into sub-seabed geological formations using pipelines originating from land is exempt. This is because the definition of 'sea' in the LP does not include the sub-seabed which is accessed only from land.[113] Importantly, the 2006 amendment to the LP lists the injection of CO_2 into geological formations in the sub-seabed in Annex I, whereas in the LC the disposal of all wastes that are not listed in Annexes I and II requires a general permit for the activity, which is not meant to be subject to much regulation.[114] This means that states seeking to inject CO_2 into the deep ocean floor (as opposed to the sub-seabed) require only a general permit under the LC. In contrast this activity is prohibited for states party to the LP because it is not listed as a permitted activity in Annex I and it is not exempt under any other provisions of the agreement.

The difference in approach means that for wastes listed in Annex I to the LP, a state must issue permits for their disposal. It must also have processes in place to support the assessment of applications using the Waste Assessment Framework (WAF) which is found in Annex II to the LP.[115] According to the LC, which requires the issue of a general permit,

---

111 Resolution LP.1(1) in Twenty-Eighth Consultative Meeting of Contracting Parties to the 1972 Convention on the Prevention of Marine Pollution by Dumping of Wastes and other Matter (October 2006) LC 28/15 and 1st Meeting of Contracting Parties to the 1996 Protocol to the Convention on the Prevention of Marine Pollution by Dumping of Wastes and Other Matter 1972 (October–November 2006) LP 1/1 ('Resolution LP.1(1) (2006)').

112 LP, opened for signature 7 November 1996, (1997) 36 ILM 1 (entered into force 24 March 2006) Art 1.4.2.1.

113 Ibid, Art 1.7.

114 LC, opened for signature 29 December 1972, 1046 UNTS 120 (entered into force 30 August 1975) Art 4(1)(c).

115 The drafting history reveals that the inclusion of the 'precautionary approach' in Art 3.1 of the LP is directly related to the introduction of 'reverse listing' and the 'waste assessment framework': see, for instance, LC/SM 1/4, annex at 10.

a state is not compelled to follow through with any of the risk management strategies introduced by Annex II to the LP. This means that the LP imposes an additional requirement that states assess the suitability of waste disposal at sea depending on the type and characteristics of the materials listed in Annex 1. This significant difference in approach means that a state under the LP cannot simply grant permits without having processes in place to assess applications taking the following into account:

- reduce waste production;
- give consideration to managing waste;
- determine the chemical, physical and biological properties of waste;
- develop an 'action list' for screening wastes in terms of their impact on human health and the marine environment;
- dump-site selection; and
- monitoring.

At present, the Scientific Group of the LP is developing specific guidance on the application of its Annex II to the states granting permits for the capture and storage of CO_2 in sub-seabed geological formations.[116] These guidelines will reflect the general framework established in Annex II but with application to sequestered CO_2.

Aside from the 'reverse listing' differences to be found between the LP and the LC, the amendment to the LP in Resolution LP.1(1)(2006) classifies the act of injecting CO_2 in repositories as 'disposal' rather than 'storage'.[117] There are different conceptual possibilities for dealing with the sequestration of CO_2 but the amendment adopts only the language of referring to the process as 'disposal'. In classifying the injection of CO_2 as 'disposal' rather than 'placement' the amendment avoids giving legitimacy to the activities of states which are members only of the LC, and which could interpret injecting CO_2 in geological formations as 'storage'. The LC, in the alternative, defines dumping in a way that excludes human activities linked to storing wastes in the seabed or subsoil, and as a result does not apply to CO_2 sequestration when states could argue that they are storing the material. The LP classes the storage of waste in the seabed and its subsoil as disposal, which is therefore prohibited according to Article IV.[118] The drafting history of the LP confirms this by showing that it was

---

116 International Maritime Organisation, 'Notification of Amendments to Annex 1 to the London Protocol 1996', LC-LP.1/Circ.5, 27 November 2006.

117 Resolution LP.1(1), above n 111.

118 LP, opened for signature 7 November 1996, (1997) 36 ILM 1 (entered into force 24 March 2006) Art 1.4.1.3.

trying to address excess gas production in offshore wells, and in particular 'the need to avoid an inadvertent prohibition of this practice'.[119]

The suggestion that injecting sequestered CO_2 into sub-seabed geological formations is storage is problematic because of the uncertainty of what this means in the context of the practice. Germany, for instance, argued during the Consultative Meetings that to mitigate climate change, CO_2 sequestration in geological formations would require that the waste be stored for more than 1000 years.[120] Good site selection for the disposal or storage of captured CO_2 is important, and if done properly not only does it reduce the risks of polluting the marine environment but it can store the material for upwards of millions of years, thereby properly mitigating climate change.[121] Whether injection of CO_2 into geological formations is intended for storage or disposal appears to be more about the certainty involved in the security of site selection rather than whether the disposal is for 1000 or 5000 years.

8.4.2 Marine Areas Regulated Through the Ocean Dumping Regime

The LP defines the term 'sea' in a broader way than the LC to include the seabed and subsoil in addition to the marine waters above it.[122] Article 1.7 of the LP merely confirms and validates earlier decisions of the Consultative Meetings to include disposal at sea of wastes under the sea-bed as part of the remit of the LC.[123] It also confirms the decision of the twelfth Consultative Meeting in 1989 that disposal into repositories

---

119 Seventeenth Consultative Meeting of Contracting Parties to the Convention on the Prevention of Marine Pollution by Dumping of Wastes and Other Matter, LC 17/14, 3–7 October 1994 [3.25].

120 Comments from Germany, 'Sequestration of Carbon Dioxide in Sub-Seabed Geological Structures: Compatibility with the London Convention/ Protocol: Legal Issues' in Submission by the United Kingdom, 'Sequestration of Carbon Dioxide in Sub-Seabed geological Structures: Compatibility with the London Convention and Protocol: Legal Issues' in Twenty-Seventh Consultative Meeting of Contracting Parties to the Convention on the prevention of Marine Pollution by Dumping of Wastes and Other Matter 1972, LC 27/INF.4, 24–28 October 2005, agenda item 6.

121 Note by the Secretariat, 'Continuation of a Review of the Benefits of Carbon Dioxide Sequestration in Sub-Seabed Geological Structures and of New Information Relevant to the Protection of the Marine Environment' in Meeting of the SG Intercessional Technical Working Group on Carbon Dioxide Sequestration, 3–7 April 2006, LC/SG-carbon dioxide 1/INF.2, agenda item 2.

122 LP, opened for signature 7 November 1996, (1997) 36 ILM 1 (entered into force 24 March 2006) Art 1.7.

123 LC 1972 Amendment Group (May 1994) LC/AM 2/4/5, [2.1].

accessed from land would not constitute dumping,[124] by specifically stating that sea does not include sub-seabed repositories accessed only from land.

The storage of CO_2 generated on land and injected into a sub-seabed geological facility using a pipeline is not dealt with by the LP. One reason given for this is that disposals from land are not 'undertaken from any structure in "marine waters" and therefore do not constitute dumping within the meaning of the London Dumping Convention'.[125] The broad definition leaves states to define for themselves whether they will regulate the use of pipelines from land to inject CO_2 directly into geological formations. It is generally accepted that the LC does not regulate this kind of activity.

8.5 ROLE AND FUNCTION OF ENVIRONMENTAL PRINCIPLES

The capture and injection of CO_2 into sub-seabed geological formations appears on its face to be of relatively low risk when compared to the potential harm from ocean acidification, resulting from the absorption of CO_2 by the ocean. It appears safer than the direct release of CO_2 into the 'ocean water column or onto the deep seafloor'.[126] What is important is that there is a degree of certainty that CO_2 will leak into the marine environment once it is injected into sub-seabed geological formations.[127] In a report published by the OSPAR Commission for the Protection of the Marine Environment of the North-East Atlantic, it is noted that 'in appropriately selected and managed geological reservoirs, the fraction' of CO_2 which is 'retained is very likely to exceed 99 per cent over 100 years and is likely to exceed 99 per cent over 1000 years'.[128] What is uncertain, however, is the impact that impurities will have on the CO_2 injected into deep-seabed geological formations. Lastly, as highlighted by this OSPAR comment, states have to regulate a number of matters to ensure the safe storage or disposal of captured CO_2.

---

124 Ad Hoc Group of Legal Experts on Dumping, 'Report of the Fourth Meeting of the Ad Hoc Group of Legal Experts on Dumping', LDC/LG 4/WP.2/Rev.1, 22–26 October 1990, [3.1.3].

125 Ibid, agenda item 7, 6.

126 Intergovernmental Panel on Climate Change, *Carbon Dioxide Capture and Storage* (2005) 3.

127 OSPAR Commission for the Protection of the Marine Environment of the North-East Atlantic, above 108.

128 Ibid, 6 (citations omitted).

How states have politically managed the risk associated with this kind of activity can be understood only in the context of social learning or the lifeworld of the international ocean dumping regime. In this sense, as discussed in section 8.3, there are significant differences between the LC and the LP, in that the focus of the latter is on eliminating and/or managing permitted disposal activities so that either they do not happen again or their effects on the marine environment are minimal. The strength of this difference resonates with the contracting parties having adopted the precautionary approach in Resolution LDC 43(14), and more formally in the LP itself and in its Article III(1) (and implicitly in other articles of the agreement).[129]

Given this significant difference, the approach of the contracting parties during negotiations to managing CO_2 capture and disposal highlights the impact that the precautionary principle has had on learning amongst states. In managing the interests of the future generation, particularly in relation to risks associated with disposing of sequestered CO_2, the precautionary principle gave creative impetus to and also constrained arguments that were acceptable during negotiations. This was possible because there was a lifeworld established by the contracting parties to the LC and the LP which defined intersubjectively how they viewed marine pollution and the political management of risk within that regime. That is, the triadic nature of negotiations is apparent in that the response of contracting parties can only be understood in the context of the social learning they have gone through over the years in developing the LP.

8.5.1 Giving Arguments a Fresh Impetus

During the negotiations of the LP, contracting parties rejected the idea that states should take a precautionary approach to situations that 'might' cause harm.[130] Article III(1) instead favoured arguments that required states to manage risks that were 'likely' to eventuate. States do not have to judge how severe the pollution is likely to be from disposal into the sea.[131] In other words, disposal activities do not have to give rise to serious

---

[129] It should be pointed out that this does not mean that only the precautionary approach is responsible for this difference.

[130] Proposal by Sweden, above n 98.

[131] The precautionary approach or principle can be expressed in many ways: see, for instance, André Nollkaemper, '"What you risk reveals what you value", and Other Dilemmas Encountered in the Legal Assaults on Risks' in David Freestone and Ellen Hey (eds), *The Precautionary Principle and International Law* (1995); Nicolas de Sadeleer, 'The Effect of Uncertainty on the Threshold Levels to

or significant grounds for concern.[132] States only need to be satisfied that a disposal activity is 'likely' to cause harm if it is permitted to go ahead. This difference is demonstrated in the LP's 'reverse listing' of materials for disposal at sea, which presumes that everything dumped at sea may or is 'likely' to cause harm irrespective of how serious it may be for the marine environment. What this does is to place an extraordinary onus on states when considering whether to introduce resolutions for amending the 'reverse list' in Annex 1. In fact, since the LP was drafted the only amendment has been in relation to the disposal of CO_2 in sub-seabed geological formations.

This middle course approach to precaution[133] appears to have given states a creative impetus to favour arguments protecting the environment more generally rather than simply reducing pollution caused by dumping of wastes at sea. That is, unlike the LC which concerns itself with weighing the risks of one kind of ocean disposal of waste with another, the LP manages waste and its impact on the natural environment more generally than just the marine environment. In the context of the capture and disposal of CO_2 the contracting parties were not just weighing the risks and potential effects of this activity in isolation from the wider problems posed by 'elevated atmospheric levels of carbon dioxide'.[134]

Contracting parties could have easily rejected the disposal of CO_2 into the sub-seabed because of the likelihood of the risk of leakage from repositories. This would have been particularly problematic if the CO_2 had contained significant amounts and types of impurities. In this case however, the precautionary approach gave creative impetus to viewing this risk as being less than the harm to the marine environment if states did not seek to reduce ocean acidification from the natural absorption of CO_2 by sea. That is, the focus of the contracting parties was shifted from the prevention of marine pollution from dumping alone to broader considerations of the likelihood of damage to the capacity of the ocean if states were not allowed to store CO_2 safely into sub-seabed repositories. Further, the

---

which the Precautionary Principle Appears to be Subject' in Maurice Sheridan and Luc Lavrysen (eds), *Environmental Law Principles in Practice* (2002) 32.

132 See, for instance, the *United Nations Framework Convention on Climate Change,* opened for signature 9 May 1992, 1771 UNTS 107 (entered into force 21 March 1994), principle 3.3 (entered into force 21 March 1994), which requires that states act in situations where there are 'threats of serious or irreversible damage'.

133 The term is developed by Nicolas de Sadeleer, *Environmental Principles: From Political Slogans to Legal Rules* (2005) 160.

134 OSPAR Commission for the Protection of the Marine Environment of the North-East Atlantic, above n 108, 1.

resolution amending Annex I to the LP also helps develop the expertise that will favour environmentally safer methods of disposing of CO_2 into the ocean than for instance its injection into the ocean floor or the marine columns above it.[135]

Importantly, the resolution amending Annex I permits disposal only of streams that 'consist overwhelmingly of carbon dioxide'.[136] Captured CO_2 can contain a host of other substances (some of which could be intentionally introduced) which could affect both the 'transport and storage' of it.[137] Given the potential 'health, safety and environmental impacts' of impure captured CO_2 streams, this requirement confirms the lifeworld established since the adoption of Resolution LDC 43(14) encouraging a precautionary approach.[138] That is, the risk associated with the activity is dealt with by requiring states to manage the purity of the captured CO_2 streams which are to be injected into sub-seabed geological formations. Further, as discussed above,[139] states must, in accordance with Annex II to the LP, use the framework on risk and waste management assessment to assess all disposal activities in its waters. States resolved to develop a specific framework for the assessment of captured CO_2 in line with Annex II to the LP.[140] This means that not only must the permit given for the injection of CO_2 deal with the purity of it but it also must deal with a host of other factors that will avert the risks associated with its disposal.

The above discussion shows that regulating what is done in the subseabed, as part of efforts to protect the marine environment itself, is a creative measure that is persuasively integrated into the ocean dumping regime. It is also a precautionary approach to protecting the water as opposed to caring for how the seabed itself is dealt with. Neither the LP nor the LC is particularly decisive in terms of whether they extend to geological formations that are thousands of kilometres below sea

---

135 On this see Intergovernmental Panel on Climate Change, above n 126.

136 Resolution LP.1(1), above n 111.

137 See the draft of the 'Risk Assessment and Management Framework for Carbon Dioxide Sequestration in Sub-Seabed Geological Structures (CS-SSGS)' which will complement Annex 2 to the 1996 Protocol: Meeting of the Scientific Group Intersessional Technical Working Group on Carbon Dioxide Sequestration, 'Report of the Meeting of the SG Intersessional Technical Working Group on Carbon Dioxide Sequestration', 3–7 April 2006, LC/SG-Carbon Dioxide 1/7, annex 3, 11.

138 Ibid.

139 See section 8.4.1 of this book.

140 For a draft of the 'Risk Assessment and Management Framework for Carbon Dioxide Sequestration in Sub-Seabed Geological Structures (CS-SSGS)' which will complement Annex 2 of the 1996 Protocol see above n 137.

level.[141] The resolution listing the disposal of CO_2 into geological formations in the sub-seabed decisively extends the regime to protect the marine waters from what is done in the seabed. [142]

The uncertainty of the potential harm to the marine environment by disposal activities as well as the impacts on climate change and acidification of the ocean if CO_2 is not sequestered and injected into the sub-seabed required a precautionary approach by states. Parties to the LP appear to have received a creative impulse from this to amend Annex I in a way that is reflective of social learning simply because the same measures would not have been possible through the LC alone.

8.5.2 The Constraining Influence of the Precautionary Principle

One of the important functions or roles played by environmental principles in their interplay with arguing during the conferences of parties of the LC and LP was to constrain how actors might have 'characterised' certain initiatives and activities in terms of whether they were to be exempt from the prohibition to dump wastes at sea. [143] In other words the precautionary principle, in the examples which will be discussed below, would have constrained how actors argued and characterised certain activities as less harmful than others and therefore allowed under the regime. It did not necessarily 'determine the particular substantive choices of the parties'

---

[141] See in particular discussions in Twenty-Seventh Consultative Meeting of Contracting Parties to the Convention on the prevention of Marine Pollution by Dumping of Wastes and Other Matter 1972, 24–28 October 2005 Agenda item 6, LC 27/6 Annex, 18 pts 102–105.

[142] Submission by the Netherlands, 'Explanatory note accompanying the proposal for restructuring and amending the London Convention' in Note by the Secretariat, 'Inventory of Amendment Proposals' in Amendment Group – First Meeting, LC/AM 1/3, 10 June 1993, agenda item 3, 42.

[143] This particular way of interpreting the constraining nature of norms is taken from Friedrich Kratochwil's characterisation of them as constitutive of what actors do: see Friedrich Kratochwil, 'How do Norms Matter?' in Michael Byers (ed), *The Role of Law in International Politics: Essays in International Relations and International Law* (2000) 34, 47. For instance, he writes at 47 that:

> [i]n so far as they are "constraints" at all, it is only in the sense identified above, in that they allow us to characterize an action as falling within a certain class, such as the making of a contract. However, these rules will not determine the particular substantive choices of the parties.

On the difference between constitutive and regulatory norms also see Nina Tannenwald, 'The Nuclear Taboo: The United States and the Normative Basis of Nuclear Non-Use' (1999) 53(3) *International Organization* 433.

but gave meaning to how they interpreted the arguments that were being presented to them, and therefore the ultimate choices and decisions that they made.[144]

This is evident for instance in the range of alternative options to the injection of CO_2 into the sub-seabed that were canvassed in reports and studies before the contracting parties to the LP and LC adopted Resolution LP.1(1)(2006). The Secretariat of the Consultative Meeting to the LC first supported a study in 1999, which it presented to contracting parties, from the Greenhouse Gas Programme of the International Energy Agency of the Organisation for Economic Cooperation and Development (OECD) relating to the idea of injecting CO_2 into 'oceanic deepwaters'.[145] In 2005 in another study prepared by the Intergovernmental Panel on Climate Change (IPCC) which was presented to the contracting parties, the potential to dispose of CO_2 in the water column and/or the sea floor was highlighted, with a caution that the ecological impact of such a venture was still in the research phase.[146] The idea of simply injecting CO_2 into the water column was at the early stages of the debate on the topic seen as something that would eventually happen anyway because of the atmosphere-ocean exchange process that naturally absorbed CO_2 into the ocean.[147] It is clear that the resolution permitting the injection of CO_2 only into sub-seabed geological formations implicitly rejected any alternative approaches to disposing of sequestered CO_2. The constraining potential of the precautionary principle is seen in the way that the regime made an issue out of the potential for leakage while the CO_2 is being transported to the storage site or in the long term after it has been stored. This suggests that gradually contracting parties narrowed their choices in terms of what they should be allowed to do with sequestered CO_2 and how much they would tolerate in terms of the risk to the marine environment. As a result, arguments

---

144 Kratochwil, above n 143, 47.

145 Note by the Secretariat, above n 107, agenda item 5.

146 Intergovernmental Panel on Climate Change, above n 126, 7. On this possibility see also support from the OSPAR Commission for the Protection of the Marine Environment of the North-East Atlantic, 'Effects on the Marine Environment of Ocean Acidification Resulting from Elevated Levels of Carbon Dioxide in the Atmosphere', in Note by the Secretariat, 'Continuation of a Review of the Benefits of Carbon Dioxide Sequestration in Sub-Seabed Geological Structures and of New Information Relevant to the Protection of the Marine Environment', Meeting of the SG Intercessional Technical Working Group on Carbon Dioxide Sequestration, 3–6 April 2006, LC/SG-Carbon Dioxide.1/INF.2, Annex at 6; also available at www.ospar.org/eng/html/welcome.html.

147 Note by the Secretariat, above n 107.

seeking to adopt alternatives to injecting CO_2 into the sub-seabed were constrained by what contracting parties had come to learn and appreciate as the level of risk they were willing to cope with in terms of pollution from the intentional disposal of CO_2 into the marine waters.[148] Contracting parties did not state that their choice to inject CO_2 into the sub-seabed, in contrast to other alternatives, was driven by the precautionary principle. However, by listing the processes of injecting CO_2 as a permissible disposal activity under Annex I to the LP they nevertheless characterised it as such.

Another example of the constraining potential of the precautionary principle is evident from the way that contracting parties approached the issue relating to whether the injection of CO_2 into the sub-seabed was to be classed as 'placement' since it is being stored or 'dumped' because it is being disposed of. The constraining potential of the precautionary approach worked to narrow the meaning that contracting parties gave to the term 'placement' in Article 2.2 of the LP.[149] To have interpreted the term 'placement' in broad terms would have meant that contracting parties were willing to allow for the possibility of CO_2 leaking from sub-seabed repositories by not regulating the activity in anyway.[150] That is, if the injection of CO_2 into the sub-seabed is interpreted as storing it then it

---

[148] Their concern with risk was also confirmed in a statement that the contracting parties to the LC and LP issued in 2007, and confirmed again in 2008, identifying their concerns about the idea of states engaging in large-scale 'fertilization of ocean waters using micro-nutrients such as iron to stimulate phytoplankton growth in order to sequester carbon dioxide': LC and LP Circular, 'Statement of Concern Regarding Iron Fertilization of the Oceans to Sequester Carbon Dioxide', LC-LP.1/Circ.14, 13 July 2007; LC and LP Circular, 'Preparation for the Scientific and Legal Discussions in 2008 of Ocean Fertilization with a View to Its Regulation Under the London Convention and Protocol', LC-LP.1/Circ.20, 29 January 2008. In particular, their statement of concerns drew from the work of the IPCC which had noted that 'ocean iron fertilization remains largely speculative, and many of the environmental side effects have yet to be assessed': Intergovernmental Panel on Climate Change, above n 126. Although not rejecting the option as yet, the contracting parties to the LC and LP have requested states to consider carefully whether to engage in such an activity and whether it would be contrary to the aims of the LC and the LP. In January the LC and LP Circular which was sent out to member states asked contracting parties to 'use the utmost caution when considering proposals for large-scale ocean fertilization operations at this stage': LC and LP Circular, 'Preparation for the Scientific and Legal Discussions in 2008 of Ocean Fertilization with a View to Its Regulation Under the London Convention and Protocol', [12.1].

[149] See above section 8.4.1.

[150] On this see in particular Submission by the United Kingdom, above n 120, annex at 11 [65].

is likely to be classified as 'placement', in which case the LP regime does not regulate the activity.

The idea that the injection of CO_2 could potentially be interpreted as 'placing' it in sub-seabed geological formations was difficult to sustain because of the requirement that the activity had to be consistent with the objectives of the LP. Arguments developed during negotiations, suggesting that injecting CO_2 into sub-seabed geological formations would protect the oceans from acidification, were seen as not consistent with the need to protect the marine environment from 'all sources of pollution'.[151] That is, the injection of CO_2 had risks for the marine environment, and an interpretation of the LP which ignored this could not have been socially acceptable. According to the scientific reports distributed through circulars to the contracting parties there was a fair amount of uncertainty as to what would make an effective site for storage and whether there would be any leakage from it or not.[152] Contracting parties did not appear to be convinced by arguments that the uncertainty of potential leakage was enough to classify the injection of CO_2 into sub-seabed geological formations as 'placement', and therefore outside the regime.

The constraining potential of the precautionary principle on what the parties accepted as good arguments seemed to have also extended to an alternative suggestion. Some states had argued that they should be allowed to 'place' CO_2 in sub-seabed repositories if they could provide a statement which illustrated how their practices were consistent with the objective of the LP, which included preventing, reducing and where practicable eliminating pollution at sea.[153] This was also rejected in favour of the amendment to the LP which classified injection of CO_2 into the sub-seabed as disposal of waste and subject to the regulatory regime established for Annex I wastes.

Clearly, states could have easily avoided the disposal of sequestered

[151] Meeting of the CM Intersessional Legal and Related Issues Working Group on Carbon Dioxide Sequestration, 'Report of the Meeting of the CM Intersessional Legal and Related Issues Working Group on Sequestration', 10–12 April 2006, LC/CM-Carbon Dioxide 1/5, pt 3.6.

[152] For example, Intergovernmental Panel on Climate Change, above n 126.

[153] Meeting of the CM Intersessional Legal and Related Issues Working Group on Carbon Dioxide Sequestration, above n 151. The argument that this should be the case was first put by the United States as one of three ways to deal with sequestration of Carbon Dioxide: see Submission by the United States, 'Facilitation and/or Regulation of Carbon Dioxide Sequestration in Sub-seabed Geological Structures: Options and Implications' at the Meeting of the CM Intersessional Legal and Related Issues Working Group on Carbon Dioxide Sequestration, 10–12 April 2006, LC/CM- Carbon Dioxide 1/3/1, pt 4.

CO_2 in geological formations in the sub-seabed. They considered several options but favoured the one that prohibited the activity unless it was regulated according to the provisions of the LP. In this instance, contracting parties were cautious and sought to regulate not only states that might harm the environment because the techniques they used for transporting or injecting and storing CO_2 into the sub-seabed might cause leakage, but also those that could seal sites without any chance of CO_2 escaping into the water column within or after more than 1000 years. In the first example discussed above contracting parties had been constrained to adopt alternative ways of injecting sequestered CO_2 into the ocean. In this second example they appear to have gone further by taking precautionary measures whereby actors could inadvertently dispose of sequestered CO_2 into the deep-seabed by classifying it as storage or placement.

8.6 CONCLUSION

The above discussion has emphasised the importance of arguing as a persuasive social process which steered states within the ocean dumping regime to learn how commonly to identify with new beliefs and preferences. Social learning through Consultative Meetings of the contracting parties led to changes being made to international law and politics by Resolution LP.1(1)(2006), amending the LP to permit states to inject CO_2 into geological formations in the sub-seabed. The resolution is much more reflective of the dynamic nature of learning at the international level, given that it does not need ratification by states.

The lifeworld within the ocean dumping regime enhanced the significance of the precautionary principle during the negotiations. This meant that contracting parties were more likely to have argued and sought to understand each other and the issues before them when negotiating whether to pass Resolution LP.1(1)(2006). This is because the lifeworld helped socially to enhance the potential for what this chapter has referred to as the triadic nature of negotiations.[154] This is in contrast to situations where states purely bargain for what they want and need. This is not to suggest that states did not bargain. In fact, the United States on a number of occasions sought to argue for alternatives to amending Annex 1 so that sequestered CO_2 was not legally regulated by the LP.[155]

The precautionary approach or principle has contributed in important

---

[154] See the discussion in section 8.2.1.
[155] See the discussion in section 8.5.2.

ways to the triadic nature of negotiations within the ocean dumping regime. However, states have not used the principle's exact provisions or wording, which appears in Article 3(1) of the LP. For instance in drafting the LP itself the idea of reverse listing appeared to have been an appeal to the idea of taking precautionary measures rather than whether the alternative approach in the LC was 'likely' to cause harm to the marine environment. That is, the precautionary principle appeared to function more as an open-textured norm or as an abstraction that pointed states to an episteme, designed to require more from them than just preventing harm from occurring. This was also the case when states considered how they should regulate the disposal of sequestered CO_2. It was shown in the discussion above in sections 8.5.1 and 8.5.2 that the precautionary principle or approach had varying roles to play in the way that Resolution LP.1(1) (2006) was passed by the Consultative Meetings to the LP and LC.

In developing arguments, appeals were made to the intersubjectively understood nature of taking precautionary measure to prevent pollution. For instance, the requirement that CO_2 injected into a sub-seabed repository should contain very few impurities is part of the precautionary measures taken to ensure that leakage has little adverse impact on the ocean itself. This suggests that social learning within the regime was relatively constructed by varying appeals to the notion of taking precautionary measure. States learned that sub-seabed repositories were important as part of the overall waste management strategy that the LP had been pursuing. Not only did the precautionary approach encourage consideration of this, but it also constrained how they regulated activities that permitted its use for the disposal of CO_2. That is, sub-seabed repositories could not be used if the CO_2 being injected into geological formations had impurities which could impact on the marine waters in the long term.

The discussion in this chapter illustrates the variable significance of an environmental principle like the precautionary principle for changing international law and politics. It has not considered what relevance the principle has for states that choose to implement the LP into domestic laws. However, at the international level states appear to have been constrained within the Consultative Meetings as to what they could do with sequestered CO_2 as well as encouraged to consider its disposal into the sub-seabed as part of an overall strategy to further protect the marine environment. In this way, the precautionary principle contributed to learning and the changes that were possible to international law and politics in the area of disposing waste into the ocean.

PART IV

Conclusion

9. Global governance of the environment: the role and function of environmental principles in changing international law and politics

9.1 INTRODUCTION

There are two broad approaches to examining the role and function of environmental principles in changing international law and politics. The first explores how actors are socialised into internalising norms which influence their interests. This description of their role commonly assumes that norms are social facts that actors will consent to adopting. From this perspective, compliance with norms and an acceptance of their legitimacy are important issues. Social processes, like persuasion and influence, are relevant in this description of change because they socialise actors into adopting norms. Environmental principles can generate similar interests amongst actors who choose to adopt them. As a result actors will be more inclined to cooperate in developing rules and institutions at the international level.

A study of the role of norms in changing international law and politics using this approach takes a narrow view of how actors' interests and identities develop. That is, the significance and potential of environmental principles for change are assessed in terms of actors complying with their provisions. Alternatively, social learning examines whether and how groups of actors learn to be interested in new preferences and develop beliefs during their engagement or interactions with each other, either directly or diffusely, at the international level itself.[1] The concept of

[1] See chapter 2 of this book where the theories of international relations relating to social learning are described and developed for the purposes of this work. Also see chapter 4 where this distinction between socialisation and social learning in the context of norms is highlighted and discussed through a reading of three approaches to abstract and open-textured norms.

learning has the potential to broaden the importance of norms, because scholars can focus both on the social processes of how groups engage with each other and the actual emergence of common knowledge and culture. It also views norms as dynamic components of the learning process rather than social facts that are stable in their meaning. How actors' interests and beliefs interlock as a result of their interactions with each other is important for describing how groups can develop common or collective interests or beliefs which are encapsulated in changing rules or norms of international law and politics.

This approach is relevant for environmental law and politics in particular because the nature of the difficulties that it seeks to address in areas such as ozone depletion and climate change all require that actors work towards sharing a common or collective view about the problems and solutions to them. A broader perspective on how norms like environmental principles function or the role they play in the process of change in international law and politics is better able to identify their significance for cooperative initiatives at the international level where some of these issues need to be addressed. This concluding chapter discusses whether and how abstract and open-textured environmental principles contribute to changing the rules and norms of international law and politics. Section 9.2 describes what it means for environmental principles to interplay with or constitute the processes that steer learning amongst groups of actors at the international level. In particular it contextualises the variability of the social processes described and examined in earlier chapters in terms of steering social learning towards changing what is common knowledge or collective culture amongst groups of actors at the international level. Section 9.3. discusses the significance of abstract and open-textured environmental principles in terms of their potential role and function through social learning to influence change in international law and politics. The last part of this section also highlights the importance of the discussion in this chapter for global governance of the environment more generally.

9.2 STEERING SOCIAL LEARNING AND CHANGE IN THE NORMS OF INTERNATIONAL LAW AND POLITICS

Through a variety of processes such as the exercise of diffuse power, modelling and emulating others, and argumentation during negotiation persuasion and social influence take place in the context of groups of actors interacting and engaging with each other. What is important about these social processes is that they explain how actors intersubjectively connect by interacting and engaging with others, either directly, or diffusely, without being materially coerced to adopt particular positions.[2] The distinction between genuine interaction and engagement generating intersubjectivity and material influences or coercion in bringing actors to the point of agreement with others is important. These social processes describe the potential to steer a group of actors to connect intersubjectively in a way in which they identify with each other about what is appropriate in a particular context, or more generally in terms of what their preferences or beliefs should be in relation to something in particular. However, these social processes can steer learning in both direct and diffuse ways, in contrast to coercion which applies causally through, for instance, the direct use of economic or military power.

This book has argued that the International Court of Justice (ICJ) has learnt to identify with a unique understanding of what is Principle 2 of the *United Nations Declaration on Environment and Development*,[3] or Principle 21 of the 1972 *Declaration of the United Nations Conference on the Human Environment*.[4] They have continued to assert their view of the norm despite the Court's subsidiary role in the processes of creating international law.[5] If the material or structurally coercive pressure of what their role should be as a particular type of court in the international law system had been influential, the group of judges on the bench might not have engaged in developing their common understanding of their particular interpretation of the transboundary harm principle. This example highlights the social nature of learning amongst groups as opposed to other coercive pressures that could potentially influence the development of intersubjectivity.

In a different example, in meetings of contracting parties to a multilateral agreement there are opportunities for states to negotiate with each other, not just to assert their interests but also to better understand the reaction of others to their ideas and preferences. More importantly their

---

2 Trine Flockhart, '"Complex Socialization": A Framework for the study of State Socialization' (2006) 12(1) *European Journal of International Relations* 89. See also chapter 2 of this book, and section 2.3. in particular, for a more detailed explanation and description of these ideas.

3 UN Doc A/CONF.151/5/Rev.1 (1992) ('*Rio Declaration*').

4 UN Doc A/CONF.48/14/Rev.1 (1973) ('*Stockholm Declaration*'). This principle is hereinafter referred to as the 'transboundary harm principle'.

5 Art 38(1)(d) of the *Statute of the International Court of Justice* states that judicial decisions are a 'subsidiary means for the determination of the rule of law'. Also, Art 59 says that '[t]he decision of the Court has no binding force except between the parties and in respect of that particular case'. For a discussion of these points and the role of the ICJ internationally see chapter 6 of this book, and in particular section 6.2.

engagement with others enables them to understand what is appropriate or common sense, and also potentially to be persuaded during their formal and informal engagements in those meetings. This seems apparent for instance in the context of the Consultative Meetings of the contracting parties to the 1972 *Convention on the Prevention of Marine Pollution by Dumping of Wastes and Other Materials*,[6] and the 1996 *Protocol to the Convention on the Prevention of Marine Pollution by Dumping of Wastes and Other Matter*,[7] that passed Resolution LDC.44(14) (1991) requiring that states issue licences for the injection of carbon dioxide (CO_2) into geological formations in the sub-seabed.[8] The injection of CO_2 into the sea and the sub-seabed is increasingly a possibility for many states. At a time when the international community is concerned with the impact of increasing CO_2 absorption by the ocean, the regime was fairly proactive in prohibiting its indiscriminate disposal at sea. It adopted only an exception which allows the injection of CO_2 into the sub-seabed. This measure is designed to protect the marine environment further from the natural absorption of CO_2 by the ocean waters. This example highlights the potential that argumentation had in the context of the Consultative Meetings of contracting parties to the LC and the LP to steer what the group of actors learnt in relation to the disposal of wastes at sea.

The interlocking beliefs of groups of actors were shaped by their negotiations which had a persuasive impact on those who might have approached the activity with different interests and beliefs about the injection of CO_2 into the ocean waters and sub-seabed. Social learning in the context of the above example is seen through the emergence of the common understanding which took the form of Resolution LDC.44(14) (1991) amending its Annex 1 to allow the disposal of CO_2 only with a licence. There are other possibilities of this kind of argument in the context

---

[6] Opened for signature 29 December 1972, 1046 UNTS 120 (entered into force 30 August 1975) ('LC'). The agreement was renamed in 1993 as the *London Convention.*

[7] Opened for signature 7 November 1996, (1997) 36 ILM 1 (entered into force 24 March 2006) ('LP'). The term ocean dumping regime will be used to refer to the collective impact of the LC, LP, and the resolutions of the consultative committee to both agreements. The LP as of 31 March 2008 had 34 contracting parties compared to the 82 that had signed and ratified the LC. See the official website of the International Maritime Organisation, www.imo.org/ accessed 1 December 2009.

[8] LDC 44(14), in Fourteenth Consultative Meeting of Contracting Parties to the Convention on the Prevention of Marine Pollution by Dumping of Wastes and Other Matter, LC 14/16/Annex 2, 30 December 1991.

of international institutions in particular.[9] This can also be observed in the way that the ICJ has relied on the potential for dialogue and negotiations amongst states after adjudicating on their dispute.[10] In a number of cases the ICJ has guided the disputing parties before it to consider certain matters in the light of the decision that it has handed down on certain aspects of their rights and liabilities towards each other. The diffuse power of the ICJ is communicated to the parties through its decision, but also the manner in which it guides those using norms like environmental principles. In this way parties that negotiate with each other following a dispute will be engaging in diplomatic relations of the kind necessary for creating common knowledge which could be codified in subsequent agreements that they have drafted between them. This type of social learning is also significant for international law and politics, given the potential that bilateral disputes have to escalate and remain as persistent strains in the diplomatic relationship between states for protracted lengths of time.

Common knowledge in the form of interlocking ideas and beliefs about what is appropriate is not however the only way that change in the rules and norms of international law and politics can be steered socially. The Global Compact (GC) initiative of the United Nations (UN), for instance, by establishing a framework that supports and encourages the social diffusion of ideas and innovations, provides opportunities for corporations to adopt practices that are consistent with their corporate strategies.[11] In this sense diffusion supports the socialisation of corporations into recognising that certain ideas and innovations have value for them even in terms of enhancing their business strategies. Given the current cultural bias towards unrestrained patterns of production and consumption, driven strongly by corporations from around the globe, a cultural shift in any form favouring environmental stewardship will be significant for what is possible in the future. As a result the instantiation of a cultural and political norm which takes more seriously attempts of transnational corporations at environmental stewardship is collective learning in the long term. Importantly, the difficult history that the UN has had in directly regulating corporations of any kind means that a collective culture of engaging with corporations at the international level is in itself significant for future regulatory measures through norms of international law and politics.

---

9 See, for instance, Peter Haas, 'International Institutions and Social Learning in the Management of Global Environmental Risks' (2000) 28(3) *Policy Studies Journal* 558.

10 See chapter 6 of this book for an examination of this argument.

11 See chapter 7 of this book for an examination of this argument.

9.3 SOCIAL LEARNING AND ENVIRONMENTAL PRINCIPLES

To steer social learning amongst groups of actors, ideas and norms are important in a variety of ways. However, the manner in which ideas and episteme are abbreviated by environmental principles or the open-textured way that they are structured as norms is also important. This is because principles' interplay with the dominant social processes within the interactions and engagements of groups of actors creates meaning for the way in which change in terms of the norms of international law and politics takes place. More crucially, it is possible to argue that certain types of learning amongst groups of actors would not happen were it not for the abstract or open-textured nature of environmental principles.

9.3.1 The 'Weight' of Environmental Principles and their Role in Changing the Norms of International Law and Politics

The importance of this interplay between environmental principles and processes that steer the interactions and engagements of actors to learn means that certain issues that normally plague abstractly stated or open-textured norms are not as significant. For instance the weight that the international community associates with certain environmental principles is not as important as whether they are likely to interplay with or constitute particular social processes.[12] In other words, there are potentially a significant number of norms that can equally claim to be environmental principles because they are abstract and open-textured. Whether or not groups are willing to create meaning from them will depend on the nature of the lifeworld[13] or types of episteme that institutions can create,[14] and which determine the extent to which they will interplay with or constitute processes necessary for social learning.[15] For instance, in the ocean dumping

---

[12] This is an obvious issue for Dworkin, who distinguishes rules from environmental principles on the basis that the latter have a dimension of weight which rules do not. He says that '[p]rinciples have a dimension that rules do not – the dimension of weight or importance': Ronald Dworkin, *Taking Rights Seriously* (1977) 26.

[13] See chapter 8, and section 8.2 in particular, for a discussion of this concept in the context of arguing and negotiation.

[14] See chapter 6, and section 6.4 in particular, for a consideration of this idea in the context of post-adjudication dialogue amongst states.

[15] This is arguably another way of making a point raised by Freestone, who has written that it 'does not mean that the application of a principle is automatic in every case nor indeed that the method of its application is the same, or even

regime established initially by the LC and later built on by the LP, the precautionary principle was far more active in the learning process as a norm than the polluter pays principle.[16] This is most apparent in relation to how the contracting parties dealt with the injection of CO_2 into the ocean and its seabed, where consideration of the polluter pays principle hardly came up in the reports of their meetings and in documents supplied by states. One obvious explanation for this is that the LP, in amending how it managed waste disposal through Annex 1, had significantly departed from approaches that had been taken through the LC. The lifeworld established through the LP and its Annex 1 provides a useful explanation of why the precautionary principle constituted the nature of the argumentation relating to how the regime ultimately dealt with the disposal of CO_2 into the ocean and sub-seabed.

This discussion also differentiates this approach to social learning through environmental principles as the dynamic for changing the norms of international law and politics from those, for instance, of Brunnée and Toope's constructivist reading of learning and change in international law.[17] The learning process itself associates weight to environmental principles in the context of engaging with each other. Common knowledge that emerges from the interaction of actors with each other, when expressed as a norm of international law, need not be associated with values, such as fairness and justice, which law prioritises. The stability in the change process comes from actors being intersubjectively connected to others as a result of their interactions about what they want and believe in that particular context. In the above example, the amendment to Annex 1 and the LP's regulation of the disposal of CO_2 happened because of the lifeworld of the ocean dumping regime and the social learning that the precautionary principle has helped generate.

It is arguable that the weight and importance of environmental principles for changing the norms of international law and politics are also relative to the nature of social learning taking place through the interaction of actors with each other. For instance, the GC created the vague and open-textured environmental principle requiring that 'businesses should undertake initiatives to promote greater environmental responsibility'.[18]

similar, in every case': David Freestone, 'Caution or Precaution: "A Rose by any Other Name. . .?"' (1999) 10 *Yearbook of International Environmental Law* 25, 26.

[16] See chapter 8, and section 8.3 in particular, for a discussion of environmental principles within the ocean dumping regime.

[17] See chapter 4, and section 4.2 in particular.

[18] See United Nations Global Compact, *The Ten Principles* (2008), www.unglobalcompact.org/AboutTheGC/TheTenPrinciples/index.html. See chapter 7

The importance of this norm relative to the other two environmental principles of the GC[19] depends on the extent to which it supports the diffusion of ideas and innovations, and as far as corporations can define their type of membership in the GC through the vague and open-textured idea of 'responsibility'.[20] This is because the issue, when contextualised in terms of social learning, is not whether the precautionary principle is abstractly more or less useful in protecting the environment, but which environmental principles will engage the greatest number of members of the GC to instantiate an alternative approach to consumerism and production which has been earlier described as the social learning that is being potentially steered through the GC. Therefore there might be an argument to be made that the principle of 'environmental responsibility' is likely to engage more corporations in the social learning of a culture that in the long term will have greatest impact on the norms of international law and politics.

This discussion highlights the value of assessing the role of environmental principles, not as soft law which actors need to be socialised into adopting but in terms of the kind of social learning that they are a part of amongst a group of actors.[21] The GC example above illustrates this point because the significance of the principle of 'environmental responsibility' could easily be lost if examined in terms of whether it can adequately regulate what corporations should do in comparison to the precautionary approach, which has a more established history as a norm in international law and politics.[22] Another example which illustrates this point is the way in which sustainable development helped produce the power of the ICJ in the *Case Concerning the Gabčíkovo-Nagymaros Project* (*Hungary* v *Slovakia*),[23] over its interpretation of the concept of 'equitable utilisation' of an international watercourse.[24] From a soft law perspective the concept of sustainable development could be viewed as having done little for the

---

of this book for a discussion of environmental principles in the context of social learning and the GC.

[19] Principles 7 and 9 are respectively the precautionary approach and that 'business should encourage the development and diffusion of environmentally friendly technologies': ibid.

[20] See chapter 7 of this book, and section 7.4 and 7.5 in particular, for a discussion of these ideas.

[21] See chapter 4, and section 4.3 in particular, for a discussion of soft law in the context of social learning at the international level.

[22] See chapter 7 of this book, and section 7.2.1 in particular, for more detailed discussion of the principles of the GC relative to each other.

[23] [1997] ICJ Rep 7 ('*Danube Dam Case*').

[24] See chapter 6, and section 6.5.2 in particular, for a detailed discussion of this idea.

actual decision of the ICJ in the *Danube Dam Case*. However, the ICJ significantly valued environmental harm which the Danube and its riparian areas were exposed to as a result of the diversion by Slovakia of portions of the river. In assessing what 'equitable' meant the ICJ, as a result of its interactions in the *Danube Dam Case*, appears to have emphasised that a state cannot be deprived of its interests in the sustainable use of its resources. This aspect of the decision provides weight to the principle of sustainability as part of the learning experience of the Court in interpreting existing provisions of international law without directly applying it in the *Danube Dam Case* itself.

9.3.2 The Dynamic Nature of the Meaning of Environmental Principles and their Role in Changing International Law and Politics

Another common issue amongst those who criticise the potential role of norms that operate at the international level is that their meaning changes, and because of this they are ineffective, uncertain, and unpredictable across a period of time and contexts.[25] This is also an issue for environmental principles to the point where Beyerlin has suggested that the price which is paid by environmental agreements is high when they use environmental principles. [26] This is because 'vague norms tend to deprive agreements of any strength, unless an explicit agreement provides for dispute settlement, which could allow vague treaty norms to be given a more precise meaning through adjudication'.[27] This might be the case if one assesses their role and function in terms of socialising actors into adopting particular principles, but an approach to social learning values how actors create their own meaning from environmental principles during their direct or diffuse interactions and engagement with each other.

In fact, the abbreviated way in which environmental principles direct attention to an episteme rather than a particular narrowly defined set of normative propositions means that groups can draw from a variety of ideas in applying themselves to a particular situation. Although the open-textured nature of some environmental principles provides more concrete

---

[25] For discussion of this issue see chapter 1, especially section 1.2, and chapter 2 more generally. In the context of the concept and principle of state sovereignty see Christian Reus-Smit, 'Human Rights and the Social Construction of Sovereignty' (2001) 27 *Review of International Studies* 519.

[26] Ulrich Beyerlin, 'Different Types of Norms in International Law' in Daniel Bodansky, Jutta Brunnée, and Ellen Hey (eds), *The Oxford Handbook of International Environmental Law* 426, 428.

[27] Ibid.

direction, it cannot be presumed that the meaning of certain terms is constantly the same or certain even at a particular point in time because of the varying values or philosophical orientations that individuals will have to nature. These aspects or qualities of environmental principles highlight the potential that meaning will emerge through the learning process within groups rather than something that they will adopt because it is a social fact that has universal relevance for all groups of actors at all times. In other words, whether or not environmental principles mean the same thing across different periods of time does not say much about their determinacy within a particular context where groups socially learn to associate their own particular interests with the norm.

The use of the precautionary approach as adopted by the LP illustrates this point. The principle provided the contracting parties with the possibility to learn socially about how they should regulate the disposal of CO_2 into the sub-seabed repositories. This is the case even though the precautionary principle's definition in Article 3(1) of the LP specifies the open-textured proposition that disposal activities should be restricted if they are 'likely to cause harm even when there is no conclusive evidence to prove the causal relation between inputs and their effects'. In other words, the terms 'likely to cause harm' opens up many opportunities for indeterminacy and inconsistencies in terms of how individual actors will apply them to their own circumstances. The precautionary approach had clearly engaged the contracting parties who, during the Consultative Meetings to the LC and LP, took steps to prevent the indiscriminate disposal of CO_2 into the ocean. They did this by adopting a resolution subjecting the disposal of CO_2 into sub-seabed repositories to the regulatory scheme of Annex 1 to the LP.[28] It is arguable that there was not enough certainty that injecting CO_2 into the sub-seabed would harm the marine environment. Instead of prohibiting it because it 'may' cause harm they regulated it by requiring licences to be issued by states. In taking this step, the contracting parties defined the terms 'likely' and 'may' relative to the overall harm to the marine environment if levels of CO_2 that are naturally absorbed into the atmosphere were not reduced in some way by allowing its injection into sub-seabed repositories.

This last example is, however, different from situations where the social learning defines the intersubjectivity or common understanding that groups wish to associate with the principle itself. That is, learning designed to provide nuance and context for the application of principles in other situations where they have to engage with them again. This does not

[28] See chapter 8, and section 8.4 in particular.

mean that its meaning for the group is stabilised, but that through their engagements with each other they give it direction for future purposes. For instance, the ICJ in the *Danube Dam Case*, in an indirect way, indicated that the precautionary principle did not extend as far as legitimising 'uncertain' risk, and therefore had no relevance for its interpretation of the concept of 'imminent peril' in relation to the test of ecological necessity.[29] Importantly, the ICJ judges appeared to be influenced by the meaning the precautionary principle had for them only in terms of legitimising and preventing 'certain' risk or likely harm. In highlighting the kind of risk that the ICJ is willing to deal with it is arguable that in this case the learning process developed the kind of intersubjectivity that can be associated with the precautionary principle itself in future cases.

This is also the case in terms of the meaning of the transboundary harm principle which is clearly different from the more prominent formulations of it in international law.[30] This does not suggest that one definition is better than the other. Instead it points to the role of social learning amongst groups of actors, because common understanding or intersubjectivity is more reflective of the true meaning of the norm for the Court itself. In the case of the transboundary harm principle, for instance, the ICJ has on more than one occasion confirmed the idea that the environment which is protected through the principle is 'the living space, the quality of life and the very health of human beings, including generations unborn'.[31] In confirming its anthropocentric perspective on the norm it has also defined in broad terms the interests of the future generation that the Court seeks to protect. However, this meaning it has ascribed to the transboundary harm principle is clearly one that the bench of the ICJ subscribes to rather than something that can be universally asserted as the definition of the norm.

9.3.3 Variability, Function of Environmental Principles, and Change in the Norms of International Law and Politics

The function and role of environmental principles will vary depending on the nature of the way that environmental principles interplay with processes that steer social learning. What is common amongst the variety of ways that they influence change in the norms of international law and

---

[29] See chapter 6, and section 6.5.1 in particular, for a more detailed discussion of this idea.

[30] See chapter 6, and section 6.6 in particular.

[31] *Legality of the Threat or Use of Nuclear Weapons*, Advisory Opinion [1966] I ICJ Rep 241–242, [29].

politics is their ability to frame or structure the discourse or ideas that interplay with social processes. Five different ways that environmental principles frame and structure discourse or ideas within groups of actors are discussed below to highlight their functional potential for change in international law and politics. The effects of regulating or constituting the social interactions of actors with each other are bound to vary, and these illustrations highlight this fact. Importantly, the significance of environmental principles is in their potential role in a variety of different contexts.

9.3.3.1 Privileging particular kinds of ideas and innovations that favour protecting the environment

Environmental principles can constrain what ideas are explored, put forward or seen as being appropriate within negotiations. This was apparent for instance in the negotiations on the development of a response to the injection of CO_2 into the sub-seabed during the Consultative Meetings of the LP and LC.[32] Contracting parties were clearly constrained when amending Annex I not to allow the general disposal of CO_2 into the ocean, which was more 'likely' to harm the environment than simply injecting it into sub-seabed repositories. In an alternative case, the GCO, by supporting three environmental principles and engaging actors in a range of activities which are designed to enhance their understanding of these norms, privileges certain ideas and innovations from diffusing within the framework being established by the GC. This means, for instance, that ideas which are consistent with corporations taking 'environmental responsibility' will be privileged through the various programmes established through the GCO.

9.3.3.2 Framing opportunities for the creative engagement of actors with environmental concerns

It is arguable that were it not for the transboundary harm principle the ICJ would not have been able to commit itself to learning about how states should prevent harm because of its consequences for the future generation rather than just the current one.[33] In a different way, the negotiations within the Consultative Meetings of the LC and LP saw the advantages to regulating the injection of CO_2 into sub-seabed repositories because the risk of the ocean naturally absorbing it was more likely than its leakage

---

[32] See chapter 8, and section 8.5.1 in particular.

[33] See chapter 6, and section 6.6 in particular, for a discussion of the ICJ and transboundary harm principle.

into the ocean waters.[34] In both these instances, environmental principles highlight the novel ways that groups of actors were able to think through their decisions when interacting with others.

9.3.3.3 Socially structuring how actors will identify their environmental credentials in relation to others

Environmental principles can provide the language and ideas that groups of actors need to distinguish themselves from others as a particular type of actor. The GCO, for instance, has established and administers a process whereby actors Communicate on their Progress (COP) with other corporations and stakeholders in implementing environmental principles. In engaging with this process corporations give meaning to the type of identity that the GC aims to create amongst its participants. The ability and interest of corporations to identify themselves in a way that is socially acceptable to others observing the GC are determined by their ability to portray themselves through their implementation of environmental principles. In this way environmental principles frame the ideas that actors need to identify socially with corporations that are citizens or members of the GC framework.

9.3.3.4 Giving particular structure to the meaning that general or environmentally specific rules and norms have for groups

Environmental principles can provide the background knowledge and ideas that structure what groups believe as the common sense approach to interpreting existing norms and rules relevant to their interactions. In this way they may shape how groups think about established interpretations of their existing practices. In the *Danube Dam Case* for instance the views of the ICJ in relation to the 'equitable utilisation' of an international watercourse were arguably produced by the principle of sustainable development.[35]

9.3.3.5 A mechanism for institutions diffusely to communicate with actors who are engaging bilaterally with each other outside their direct reach

Institutions may engage with actors who choose to engage in bilateral relations outside the particular institutional context. They are able to rely on the intersubjectivity that the international community has associated with environmental principles to communicate with them and seek to influence

---

[34] See chapter 8, and section 8.5.1. in particular, for this discussion.
[35] See chapter 6, and section 6.5.2. in particular, for this argument.

their interactions in a diffuse way. The principles adopt their communicative potential by drawing on the episteme created within the institutions themselves.[36] For instance, the ICJ in the *Danube Dam Case* drew on the communicative potential of environmental principles to guide Slovakia and Hungary's post-adjudication interactions with each other.[37]

What is apparent from the above list is that the abstract or open-textured nature of environmental principles creates their significance for changing international law and politics. As a frame or structure of thought rather than as a defined and concrete way of doing something their importance is both in terms of their versatility and flexibility in terms of role and function. Importantly this versatility also means that as intersubjective structures of ideas and discourses their influence will develop as they are embedded further into the common understandings and collective cultures of groups internationally.

9.4 IMPORTANCE OF ENVIRONMENTAL PRINCIPLES FOR PROTECTING THE ENVIRONMENT

Drafting treaties or conventions to bind states formally is sometimes a laborious diplomatic process that takes long periods of time to complete. Even when agreements open for signature, states may not always find themselves interested in particular aspects of them and may withdraw from signing and ratifying them at the last minute.[38] This challenge is even more complicated in the case of setting standards for transnational corporations which can shift the forums of their operations more successfully than other transnational actors. Important gaps exist in global measures that are aimed to protect the environment because of the reluctance of states to draft international codes of conduct that will directly regulate corporate behaviour.

---

[36] On the concept of an episteme see the discussion in chapter 6, and section 6.4. in particular.

[37] See chapter 8, and section 6.4. in particular, for this argument.

[38] As an example, Rothwell and Podehl have discussed how New Zealand abandoned the 1988 *Convention on the Regulation of Antarctic Mineral Resource Activities*, opened for signature 25 November 1988, (1988) 27 ILM 868 (not yet in force) after playing a significant role during its negotiation: Jorg Podehl and Donald Rothwell, 'New Zealand and the Convention on the Regulation of Antarctic Mineral Resource Activities (CRAMRA): An unhappy divorce?' (1992) 22 *Victoria University of Wellington Law Review* 23.

Environmental principles create opportunities for a range of actors to participate within interactions that groups have with each other at the international level. This book has illustrated how, from a broader perspective on the nature of change in international law and politics, actors' interests and identities are also created as a result of their engagement with others at the international level. Environmental principles can engage a range of actors with each other because, individually or collectively, actors can create some relevant meaning from principles. As a result they can generate the necessary interests and identities amongst actors in the short or long term that will support creating new rules and regulatory mechanisms through multilateral engagements. In this sense, approaches to the governance of environmental issues could be more successful if groups of actors could initially engage with each other over periods of time, negotiating and examining issues in the light of environmental principles. For instance, the precautionary principle had helped build a degree of intersubjectivity amongst states in relation to how contracting parties to the LP might value and negotiate issues relating to risk and the uncertainty of harm associated with disposal of waste at sea. This clearly provided an easier inroad for the contracting parties to reject various practices for disposing of CO_2 into the ocean waters and instead seek to regulate the least risky strategy.

Although in the short term actors will use the vague or open-textured nature of environmental principles to adopt or value only minimal changes to their interests, it is also the ability of such social practices to instantiate a deeper cultural change in relation to established patterns. This can include consumerism and production that makes them important for protecting the environment. That is, in the long term the common sense approach to business means that corporations are ordinarily aware of the impact that their actions will have on the environment and the depletion of natural resources. The ability of environmental principles potentially to stimulate certain practices that can generate further structural pressure on transnational corporations is significant, given that their direct regulation at the international level is currently both important and difficult.

This book has explained the relative nature of ideas and concepts in constructing international environmental law and politics. It has not sought to address the source of these environmental principles, which is the subject of research by others examining norms especially from constructivist perspectives in international relations.[39] It does, however,

---

[39] See, e.g., Susan Park, 'Theorizing Norm Diffusion Within International Organizations' (2006) 43 *International Politics* 342.

point to the fact that not all discourse or ideas supported through the framework or structure of particular environmental principles are good for nature and the development of global responses to environmental harm. For instance, the transboundary harm principle discussed in the various sections of this chapter and work highlights and supports, in its original formulations in Principle 2 of the *Rio Declaration* and Principle 21 of the *Stockholm Declaration*, the sovereign rights of states over environmental stewardship. This is in contrast to concepts and principles like 'common area', 'common heritage', and 'common concern' which emphasise humanity's collective concerns with protecting the environment.[40] The differences between these principles are significant for the direction that environmental law and politics can take in the future. The following passage from Judge Weeramantry, as he was when he delivered his separate opinion in the *Danube Dam Case*, sums up well the significance of seeking to construct alternative ways of protecting the environment without being concerned with the sovereign rights of states which, for instance, the transboundary harm principle emphasises:

> We have entered an era . . . in which international law subserves not only the interests of individual States, but looks beyond them and their parochial concerns to the greater interests of humanity and planetary welfare. In addressing such problems, which transcend the individual rights and obligations of the litigating States, international law will need to look beyond procedural rules fashioned for purely *inter partes* litigation.
>
> . . .
>
> International environmental law will need to proceed beyond weighing the rights and obligations of parties within a closed compartment of individual State self-interest, unrelated to the global concerns of humanity as a whole.[41]

This passage highlights the need to critique the kind of environmental principles that support the continuing instantiation of the persistent cultural understanding that the environment and its protection are not humanity's concern, but rather subject to the sovereign will of states. It also points to the limitations inherent in the disciplinary approaches to law and norms at the international level. This means that humanity's ability to protect the environment is constrained not just by what we do not know. Weeramantry also seems to suggest that we need to rethink our capacity

---

40 Jutta Brunnée, 'Common Areas, Common heritage, and Common Concern' in Daniel Bodansky, Jutta Brunnée and Ellen Hey (eds), *The Oxford Handbook of International Environmental Law* (2007) 550, 552.

41 *Case Concerning the Gabčíkovo-Nagymaros Project (Hungary/Slovakia)* [1997] ICJ Rep 7, Weeramantry V-P at 115.

to order and change what we do at the international level so that we are concerned more with nature and its preservation for current and future generations than the immediate self-interests of states.

This book has shown how a broader approach to understanding change at the international level suggests that we do not have to be totally concerned with individual actors' interests as they exist apart from their interactions with others. It is possible to observe actors cooperating and collectively identifying with each other at the international level in achieving change, even within institutions which traditionally cater to coordinating what they want from each other. This is particularly important for initiatives relating to corporations that are global actors with private rather than public interests and which have not traditionally had much interest in taking social responsibility for the environment. This means that by broadening our understanding of the nature of normative change at the international level we can better understand the disciplinary boundaries of international law which Weeramantry critiques in the above passage. In this context, environmental principles are particularly significant because they ideologically frame the social processes which steer common and collective orientations of global actors to environmental issues.

Bibliography

Aalberts, Tanja (2004), 'The Future of Sovereignty in Multilevel Governance Europe – A Constructivist Reading', 42(1) *Journal of Common Market Studies*, 23.

Abbott, Kenneth and Duncan Snidal (2000), 'Hard and Soft Law in International Governance', 54(3) *Legalization and World Politics* 421.

Abbott, Kenneth and Duncan Snidal (1998), 'Why States Act Through Formal International Organizations', 42(1) *Journal of Conflict Resolution* 3.

Abbott, Kenneth, Robert Keohane, Andrew Moravcsik, Anne-Marie Slaughter and Duncan Snidal (2000), 'The Concept of Legalization', 54(3) *International Organization* 401.

Aceves, William (2000), 'Critical Jurisprudence and International Legal Scholarship: A Study of Equitable Distribution', 39 *Columbia Journal of Transnational Law* 299.

Adede, Andronico (1992), 'International Environmental Law from Stockholm to Rio – an Overview of Past Lessons and Future Challenges', 22(2) *Environmental Policy and Law* 88.

Adler, Emanuel (1997), 'Seizing the Middle Ground: Constructivism in World Politics', 3(3) *European Journal of International Relations* 319.

Adler, Emanuel and Steven Bernstein (2005), 'Knowledge in Power: the Epistemic Construction of Global Governance' in, Michael Barnett and Raymond Duvall (eds), *Power in Global Governance*, p.294.

Adler, Emanuel (2005), *Communitarian International Relations: The Epistemic Foundations of International Relations*.

Ahrne, Goran and Nils Brunsson (2004), 'Soft Regulation from an Organizational Perspective' in Ulrika Mörth (ed.), *Soft Law in Governance and Regulation: An Interdisciplinary Analysis*, p.171.

Åkermark, Sia Spiliopoulou (2004), 'Soft Law and International Financial Institutions – Issues of Hard and Soft Law from a Lawyer's Perspective' in Ulrika Mörth (ed.), *Soft Law in Governance and Regulation: An Interdisciplinary Analysis*, p.61.

A-Khavari, Afshin and Donald Rothwell (1998), 'The ICJ and the *Danube Dam Case*: A Missed Opportunity for International Environmental Law?', 22 *Melbourne University Law Review* 507.

Alchourrón, E. Carlos and Eugenio Bulygin (1981), 'The Expressive Concept of Norms' in, Hilpinen, Risto (ed.), *New Studies in Deontic Logic*.

Alchourrón, E. Carlos and Eugenio Bulygin (1971), *Normative systems*.

Alderson, Kai (2001), 'Making Sense of State Socialization', 27 *Review of International Studies* 415.

Alderson, Kai and Andrew Hurrell (eds) (2000), *Hedley Bull on International Society*.

Aldestam, Mona (2004), 'Soft Law and the State Aid Policy Area' in Ulrika Mörth (ed.), *Soft Law in Governance and Regulation: An Interdisciplinary Analysis*, p.11.

Alexander, Larry and Ken Kress (1996–97), 'Against Legal Principles', 82 *Iowa Law Review* 739.

Allen Roberson, Barbara (ed.) (1998), *International Society and the Development of International Relations Theory*.

Allott, Phillip (1989), *International Law and the International Revolution: Reconceiving the World*, Speech delivered at the Josephine Onoh Lecture Series, University of Hull.

Anghie, Antony and Garry Sturgess, (eds) (1998), *Legal Visions of the 21st Century: Essays in Honour of Judge Christopher Weeramantry*.

Archer, Margaret (1995), *Realist Social Theory: The Morphogenetic Approach*.

Arend, Anthony (1999), *Legal Rules and International Society*.

Atapattu, Sumudu (2001), 'Sustainable Development, Myth or Reality? A Survey of Sustainable Development under International Law and Sri Lankan Law', 14 *Georgetown International Environmental Law Review* 265.

Atienza, Manuel and Juan Ruiz Manero (1998), *A Theory of Legal Sentences*.

Backer, Larry (2006), 'Multinational Corporations, Transnational Law: The United Nations' Norms on the Responsibilities of Transnational Corporations as a Harbinger of Corporate Social Responsibility in International Law', 37 *Columbia Human Rights Law Review* 287.

Baldwin, David (2002), 'Power and International Relations' in, Walter Carlsnaes, Thomas Risse and Beth A. Simmons (eds), *The Handbook of International Relations*, p.179.

Baldwin, David (ed.) (1993), *Neorealism and Neoliberalism: The Contemporary Debate*.

Baldwin, David (1999), 'Force, Fungibility, and Influence', 8(4) *Security Studies* 173.

Baldwin, David (1993), 'Neoliberalism, Neorealism, and World Politics', in David Baldwin (ed.), *Neorealism and Neoliberalism: The Contemporary Debate* 3.

Bantekas, Ilias (2004), 'Corporate Social Responsibility in International Law', 22 *Boston University International Law Journal* 309.
Barnett, Michael and Raymond Duvall (eds) (2005), *Power in Global Governance*.
Barnett, Michael and Raymond Duvall (2005), 'Power in Global Governance' in Michael Barnett and Raymond Duvall (eds) (2005), *Power in Global Governance*, p.1.
Barnett, Michael and Raymond Duvall (2005), 'Power in International Politics', 59 *International Organization* 39.
Barnett, Michael and Martha Finnemore (1999), 'The Politics, Power, and Pathologies of International Organizations', 53(4) *International Organization* 699.
Barnett, Michael and Martha Finnemore (1999), 'The power of liberal international organizations' in Michael Barnett and Raymond Duvall (eds), *Power in Global Governance* (2005) 161.
Barnett, Michael and Martha Finnemore (1999), *Rules for the World: International Organizations in Global Politics*.
Barrett, Scott (2003), *Environment and Statecraft: The Strategy of Environmental Treaty-Making*.
Barton, David and Karin Tusting (eds) (2005), *Beyond Communities of Practice: Language, Power and Social Context*.
Bederman, David (2001), 'Constructivism, Positivism, and Empiricism in International Law', 89 *Georgetown Law Journal* 469.
Bederman, David (2001), *International Law in Antiquity*.
Beland, Daniel (2006), 'The Politics of Social Learning: Finance, Institutions, and Pension Reform in the United States and Canada', 19 *Governance* 559.
Bennett, Colin (1991), 'How States Utilize Foreign Evidence', 11(1) *Journal of Public Policy* 31.
Bennett, Colin (1997), 'Understanding Ripple Effects: The Cross-National Adoption of Policy Instruments for Bureaucratic Accountability', 10(3) *Governance* 213.
Bennett, Colin (1991), 'What Is Policy Convergence and What Causes It?', 21 *British Journal of Political Science* 215.
Benvenisti, Eyal (1996), 'Collective Action in the Utilization of Shared Freshwater: The Challenges of International Water Resources Law', 90 *The American Journal of International Law* 384.
Benvenisti, Eyal (2004), 'Customary International Law as a Judicial Tool for Promoting Efficiency' in, Eyal Benvenisti and Moshe Hirsch (eds), *The Impact of International Law on International Cooperation*, p.85.
Benvenisti, Eyal and Moshe Hirsch (eds) (2004), *The Impact of International Law on International Cooperation*.

Bernstein, Richard J. (1983), *Beyond Objectivism and Relativism: Science, Hermeneutics and Praxis*.

Beyerlin, Ulrich (2007), 'Different Types of Norms in International Law' in Bodansky, Daniel, Jutta Brunnée and Ellen Hey (eds), *The Oxford Handbook of International Environmental Law*, p.425.

Bhargava, Rajeev (1992), *Individualism in Social Science*.

Bhaskar, Roy (1986), *Scientific Realism and Human Emancipation*.

Bigge, David (2004), 'Bring on the Bluewash: a Social Constructivist Argument Against Using *Nike v Kasky* to Attack the UN Global Compact', 14 *International Legal Perspectives* 6.

Birnie, Patricia (1997), 'Twentieth-Century Marine Conservation Conventions Adaptable to Twenty-First Century Goals and Principles?: Part II', 12(3) *International Journal of Marine and Coastal Law* 488.

Birnie, Patricia and Alan Boyle (2002), *International Law and the Environment*.

Björkdahl, Annika (2002), 'Norms in International Relations: Some Conceptual and Methodological Reflections', 15(1) *Cambridge Review of International Affairs* 9.

Boaventura de Sousa Santos (2002), *Toward a New Legal Common Sense*.

Bodansky, Daniel (1993), 'The United Nations Framework Convention on Climate Change: A Commentary', 18 *Yale Journal of International Law* 451.

Bodansky, Daniel (1995), 'Customary (and Not So Customary) International Environmental Law', 3 *Indiana Journal of Global Legal Studies* 105.

Bodansky, Daniel (2004), 'Rules vs Standards in International Environmental Law', 98 *American Society of International Law Proceedings* 275.

Bodansky, Daniel, Jutta Brunnée and Ellen Hey (eds) (2007), *The Oxford Handbook of International Environmental Law*.

Bomberg, Elizabeth and John Peterson (2000), 'Policy Transfer and Europeanization: Passing the Heineken Test?', Queens Papers on Europeanization No 2/2000, Queens University Belfast, School of Politics, International Studies and Philosophy.

Bosselmann, Klaus (2004), 'Environmental Governance: A New Approach to Territorial Sovereignty' in, Goldstein, Robert (ed.), *Environmental Ethics and Law* p.293.

Bourdieu, Pierre (1977), *Outline of a Theory of Practice*.

Bourne, Charles (1997), 'The Primacy of the Principle of Equitable Utilization in the 1997 Watercourses Convention', 35 *Canadian Yearbook of International Law* 221.

Boyle, Alan (1985), 'Marine Pollution Under the Law of the Sea Convention', 79 *American Journal of International Law* 347.

Boyle, Alan (2007), 'Relationship Between International Environmental Law and Other Branches of International Law' in Daniel Bodansky, Jutta Brunnée and Ellen Hey (eds), *The Oxford Handbook of International Environmental Law* 125.

Boyle, Alan (1997), 'The *Gabčíkovo-Nagymaros Case*: New Law in Old Bottles', 8 *Yearbook of International Environmental Law* 13.

Boyle, Alan (2007), 'The Environmental Jurisprudence of the International Tribunal for the Law of the Sea', 22(3) *The International Journal of Marine and Coastal Law* 369.

Boyle, Alan and David Freestone (eds) (1999), *International Law and Sustainable Development: Past Achievements and Future Challenges*.

Braithwaite, John and Peter Drahos (2000), *Global Business Regulation*.

Bratspies, Rebecca and Russell Miller (eds) (2006), *Transboundary Harm in International Law: Lessons from the Trail Smelter Arbitration*.

Brierley, James L. (1963), *The Law of Nations: An Introduction to the International Law of Peace*.

Brownlie, Ian (2003), *Principles of Public International Law*.

Brown Weiss, Edith (1988), *In Fairness to Future Generations. International Law, Common Patrimony and Intergenerational Equity*.

Brown Weiss, Edith (ed.), (1992), *Environmental Change and International Law: New Challenges and Dimensions*.

Brown Weiss, Edith (2000), 'Opening the Door to the Environment and to Future Generations' in de Chazournes, Laurence B. and Sands, Philippe (eds), *International Law, the International Court of Justice and Nuclear Weapons* p. 338.

Brown Weiss, Edith (2000), 'Common but Differentiated Responsibilities in Perspective', 96 *American Society of International Law Proceedings* 366.

Brunnée, Jutta and Stephen Toope (1997), 'Environmental Security and Freshwater Resources: Ecosystem Regime Building', 91(1) *The American Journal of International Law* 26.

Brunnée, Jutta and Stephen Toope (2000), 'International Law and Constructivism: Elements of an Interactional Theory of International Law', 39(1) *Columbia Journal of Transnational Law* 19.

Brunnée, Jutta (2007), 'Common Areas, Common Heritage, and Common Concern' in, Daniel Bodansky, Jutta Brunnée and Ellen Hey (eds), *The Oxford Handbook of International Environmental Law*, p.550.

Bull, Hedley (1970), 'Natural Law and International Relations', 5(2) *British Journal of International Studies* 171.

Bull, Hedley (2002), *The Anarchical Society: A Study of Order in World Politics*.

Burroughs, John (1997), *The (Il)legality of Threat or Use of Nuclear Weapons: A Guide to the Historic Opinion of the International Court of Justice*.

Buzan, Barry (2004), *From International to World Society? English School Theory and the Social Structure of Globalisation.*

Byers, Michael (1999), *Custom, Power and the Power of Rules: International Relations and Customary International Law.*

Byers, Michael (ed) (2000), *The Role of Law in International Politics: Essays in International Relations and International Law.*

Calder, Fanny and Malaika Culverwell (2004), 'Following up the World Summit on Sustainable Development Commitments on Corporate Social Responsibility', Interim Report, Royal Institute on International Affairs, London.

Caldwell, Lynton Keith (1996), *International Environmental Policy.*

Calhoun, Craig, Edward LiPuma, and Moishe Postone (eds) (1993), *Bourdieu: Critical Perspectives.*

Cameron, James and Juli Abouchar (1995), 'The Status of the Precautionary Principle in International Law' in David Freestone and Ellen Hey (eds), *The Precautionary Principle and International Law: the Challenge of Implementation*, p.29.

Carlsnaes, Walter, Thomas Risse and Beth Simmons (eds) (2002), *The Handbook of International Relations.*

Carlsnaes, Walter, Risse-Kappen, Thomas and Simmons, Beth (eds) (2005), *Handbook of International Relations.*

Carty, Anthony (1991), 'Critical International Law: Recent Trends in the Theory of International Law', 2 *European Journal of International Law* 66.

Chapman, Audrey and Sage Russell (eds) (2002), *Core Obligations: Building a Framework for Economic, Social and Cultural Rights.*

Chapman, Audrey and Sage Russell (2002), 'Introduction' in Audrey Chapman and Sage Russell (eds), *Core Obligations: Building a Framework for Economic, Social and Cultural Rights*, p.1.

Charlesworth, Hilary and Christine Chinkin (2004), 'Regulatory Frameworks in International Law' in Christine Parker, Colin Scott, Nicola Lacey, and John Braithwaite (eds), *Regulating Law*, p.246.

Charney, Jonathan (1998), 'Is international law threatened by multiple international tribunals?', 271 *Recueil des Cours* 101.

Charney, Jonathan (1999), 'The Impact on the International Legal System of the Growth of International Courts and Tribunals', 31 *New York University Journal of International Law and Politics* 697.

Checkel, Jeffrey (1998), 'The Constructivist Turn in International Relations Theory', 50 *World Politics* 324.

Checkel, Jeffrey (1999), 'Norms, Institutions, and National Identity in Contemporary Europe', 43 *International Studies Quarterly* 84.

Checkel, Jeffrey (1999), 'Social Construction and Integration', 6(4) *Journal of European Public Policy* 545.

Checkel, Jeffrey (2001), 'The Europeanization of Citizenship?' in Maria Green Cowles, James Caporaso and Thomas Risse (eds), *Transforming Europe: Europeanization and Domestic Change*, 180.
Checkel, Jeffrey (2001), 'Why Comply? Social Learning and European Identity Change', 55(3) *International Organization* 553.
Checkel, Jeffrey (2003), '"Going Native" In Europe?: Theorizing Social Interaction in European Institutions', 36 *Comparative Political Studies* 209.
Checkel, Jeffrey (2005), 'International Institutions and Socialization in Europe: Introduction and Framework', 59(4) *International Organization* 801.
Checkel, Jeffrey (ed.) (2007), *International Institutions and Socialization in Europe.*
Chinkin, Christine (1989), 'The Challenge of Soft Law: Development and Change in International Law', 38(4) *The International and Comparative Law Quarterly* 850.
Chinkin, Christine (2000), 'Normative Development in the International Legal System' in, Dinah Shelton (ed.), *Commitment and Compliance: The Role of Non-Binding Norms in the International Legal Systems*, p.21.
Churchill, Robin and Vaughan Lowe (1988), *The Law of the Sea.*
Churchill, Robin and Geir Ulfstein (2000), 'Autonomous Institutional Arrangements in Multilateral Environmental Agreements: A Little-Noticed Phenomenon in International Law', 94(4) *The American Journal of International Law* 623.
Coleman, Andrew (2003), 'The International Court of Justice and Highly Political Matters', 4 *Melbourne Journal of International Law* 29.
Collier, John and Vaughan Lowe (1999), *The Settlement of Disputes in International Law: Institutions and Procedures.*
Cook, Rebecca (ed.) (1994), *Human Rights of Women: National and International Perspectives.*
Cordonier Segger, Marie-Claire and Ashfaq Khalfan and Nakhjavani, Salim (2002), *Weaving the Rules for Our Common Future: Principles, Practice and Prospects for International Sustainable Development Law.*
Cordonier Segger, Marie-Claire and Ashfaq Khalfan, Markus Gehring and Michelle Toering (2003), 'Prospects for Principles of International Sustainable Development Law after the WSSD: Common but Differentiated Responsibilities, Precaution and Participation' 12(1) *Review of European Community & International Environmental Law* 54.
Cordonier Segger, Marie-Claire and Ashfaq Khalfan (2004), *Sustainable Development: Principles, Practices and Prospects.*
Cortell, Andrew and James Davis Jr (2000), 'Understanding the Domestic Impact of International Norms: A Research Agenda', 2(1) *International Studies Review* 65.

Crawford, James (2002), *The International Law Commission's Articles on State Responsibility*.

Cronin, Bruce (2002), 'The Two Faces of the United Nations: The Tension Between Intergovernmentalism and Transnationalism', 8 *Global Governance* 53.

Cutler, A Claire (2003), *Private Power and Global Authority: Transnational Merchant Law in the Global Political Economy*.

Cutler, A Claire, Virginia Haufler and Tony Porter (eds) (1999), *Private Authority and International Affairs*.

Daly, Herman (1993), 'Sustainable Growth: An Impossibility Theorem' in Herman Daly and Kenneth Townsend (ed.), *Valuing the Earth*, p.267.

Daly, Herman and Kenneth Townsend (ed.) (1993), *Valuing the Earth*.

Dawes, Robyn, Alphons Van de Kragt and John Orbell (1988), 'Not Me or Thee but We: The Importance of Group Identity in Eliciting Cooperation in Dilemma Situations: Experimental Manipulations', 68 *Acta Psychologica* 83.

de Sadeleer, Nicolas (2002), 'The Effect of Uncertainty on the Threshold Levels to which the Precautionary Principle Appears to be Subject' in Sheridan, Maurice and Lavrysen, Luc (eds), *Environmental Law Principles in Practice*, 32.

de Sadeleer, Nicolas (2004), 'Environmental Principles, Modern and Post-modern Law' in, Richard Macrory (ed.), *Principles of European Environmental Law*, p.223.

de Sadeleer, Nicolas (2005), *Environmental Principles: From Political Slogans to Legal Rules*.

de Sousa Santos, Boaventura (1987), 'Law: A Map of Misreading. Toward a Postmodern Conception of Law', 14(3) *Journal of Law and Society* 279.

De Waart, Paul, Paul Peters and Erik Denters (eds) (1988), *International Law and Development*.

Degan, Vladimir (1997), *Sources of International Law*.

Deitelhoff, Nicole and Harald Muller (2005), 'Theoretical Paradise – Empirically Lost? Arguing with Habermas', 31 *Review of International Studies* 167.

Dell, Sidney (1989), 'The United Nations Code of Conduct on Transnational Corporations' in Johan Kaufmann (ed.), *Effective Negotiation: Case Studies in Conference Diplomacy*, p.53.

Denzau, Arthur and Douglass North (1994), 'Shared Mental Models: Ideologies and Institutions', 47 *Kyklos* 3.

DesJardins, Joseph (2005), *Environmental Ethics: An Introduction to Environmental Philosophy*, 2005.

Deva, Surya (2006), 'Global Compact: A Critique of the UN's

"Public–Private" Partnership for Promoting Corporate Citizenship', 34 *Syracuse Journal of International Law and Commerce* 107.

Deva, Surya (2006), 'Sustainable Good Governance and Corporations: An Analysis of Asymmetries', 18 *Georgetown International Environmental Law Review* 707.

Dhondt, Nele (2002), 'Environmental Law Principles and the Case Law of the Court of Justice' in, Maurice Sheridan and Luc Lavrysen (eds), *Environmental Law Principles in Practice*, p.141.

Dhondt, Nele (2003), *Integration of Environmental Protection into other EC Policies: Legal Theory and Practice.*

Diez, Thomas and Jill Steans (2005), 'A Useful Dialogue? Habermas and International Relations', 31 *Review of International Studies* 127.

DiMaggio, Paul and Walter Power (1983), 'The Iron Cage Revisited: Institutional Isomorphism and Collective Rationality in Organizational Fields', 48(2) *American Sociological Review* 147.

Dobos, Daniel (2001), 'The Necessity of Precaution: The Future of Ecological Necessity and the Precautionary Principle', 17 *Fordham Environmental Law Journal* 375.

Dryzek, John (2005), *The Politics of the Earth: Environmental Discourses.*

Dryzek, John (2007), 'Paradigms and Discourses' in Daniel Bodansky, Jutta Brunnée and Ellen Hey (eds), *The Oxford Handbook of International Environmental Law*, p.44.

Dunne, Timothy (1995), 'International Society: Theoretical Promises Fulfilled?', 30 *Cooperation and Conflict* 125.

Dunne, Timothy (1998), *Inventing International Society: A History of the English School.*

Dunne, Timothy (2001), 'New Thinking on International Society', 3(2) *British Journal of Politics and International Relations* 223.

Dunoff, Jeffrey (1994), 'Institutional Misfits: The GATT, the ICJ and Trade-Environment Disputes', 15 *Michigan Journal of International Law* 1043.

Dworkin, Ronald (1977), *Taking Rights Seriously.*

Dworkin, Ronald (1978), *Taking Rights Seriously.*

Ebbesson, Jonas (1996), *Compatibility of International and National Environmental Law.*

Eckersley, Robyn (2004), 'Soft law, hard politics, and the Climate Change Treaty' in Reus-Smit, Christian (ed.), *The Politics of International Law*, p.80.

Edwards, Paul (2001), 'Representing the Global Atmosphere: Computer Models, Data, and Knowledge about Climate Change' in Clark Miller and Paul Edwards (eds), *Changing the Atmosphere: Expert Knowledge and Environmental Governance*, p.31.

Ehrlich, Eugen (1936), *Fundamental Principles of the Sociology of Law*.

Elias, Olufemi and Chin Lim (1997), '"General Principles of Law", "Soft Law" and the Identification of International Law', 28 *Netherlands Year Book of International Law* 3.

Eliason, Leslie and Emily Goldman (2003), 'Introduction: Theoretical and Comparative Perspectives on Innovation and Diffusion' in Emily Goldman and Leslie Eliason (eds), *The Diffusion of Military Technology and Ideas*, p. 1.

Ellis, Jaye (2001), *Soft Law As Topos: The Role of Principles of Soft law in the Development of International Environmental law*, DCL Thesis, McGill University.

Endicott, Timothy (1997), 'Vagueness and Legal Theory', 3 *Legal Theory* 37

Endicott, Timothy (2000), *Vagueness in Law*.

Engström, Viljam (2002), *Realizing the Global Compact*.

Erdem, Mete (1992), 'Sustainable Utilisation of International Watercourses: a Legal Overview', 12 *Istanbul University International Law Bulletin* 41.

Eriksen, Erik Oddvar and Jarle Weigard (1997), 'Conceptualising Politics: Strategic or Communicative Action?', 20(3) *Scandinavian Political Studies* 219.

Fairclough, Norman (2003), *Analysing Discourse: Textual Analysis for Social Discourse*.

Falk, Richard, Friedrich Kratochwil and Saul Mendlovitz (1985), *International Law: A Contemporary Perspective*.

Farkas, Andrew (2001), *State Learning and International Change*.

Ferguson, Yale and Richard Mansbach (1991), 'Between Celebration and Despair: Constructive Suggestions for Future International Theory', 35 *International Studies Quarterly* 363.

Fierke, Karin and Antje Wiener (1999), 'Constructing Institutional Interests: EU and NATO Enlargement', 6(5) *Journal of European Public Policy* 721.

Finn, Daniel P. (1981), 'Ocean Disposal of Radioactive Wastes: The Obligation of International Cooperation to Protect the Marine Environment', 21 *Virginia Journal of International Law* 621.

Finnemore, Martha (1993), 'International Organizations as Teachers of Norms: the United Nations Educational, Scientific, and Cultural Organization and Science Policy', 47 *International Organization* 565.

Finnemore, Martha (1996), *National Interests in International Society*.

Finnemore, Martha (2000), 'Are Legal Norms Distinctive?', 32 *New York University Journal of International Law and Politics* 699.

Finnemore, Martha and Kathryn Sikkink (1998), 'International Norm Dynamics and Political Change', 52(4) *International Organization* 887.

Finnemore, Martha and Stephen Toope (2001), 'Alternatives to "Legalization": Richer Views of Law and Politics', 55(3) *International Organization* 743.

Fitzmaurice, Gerald (1957), 'The General Principles of International law, Considered from the Standpoint of the Rule of law', 95 *Recueil des Cours* 7.

Fitzmaurice, Malgosia and Milena Szuniewicz (eds) (2003), *Exploitation of Natural Resources in the 21st Century.*

Flockhart, Trine (2004), '"Masters and Novices": Socialization and Social Learning through the NATO Parliamentary Assembly', 18 *International Relations* 361.

Flockhart, Trine (2006), '"Complex Socialization": A Framework for the Study of State Socialization', 12(1) *European Journal of International Relations* 89.

Flynn, Gregory and Henry Farrell (1999), 'Piecing Together the Democratic Peace: The CSCE, Norms, and the "Construction" of Security in Post-Cold War Europe', 53 *International Organization* 505.

Ford, Christopher (1994), 'Judicial Discretion in International Jurisprudence: Article 38(1)(c) and "General Principles of Law"', 5 *Duke Journal of Comparative and International Law* 35.

Franck, Thomas (1975), 'World Made Law: The Decision of the ICJ in the Nuclear Tests Cases', 69 *American Journal of International Law* 612.

Franck, Thomas (1988), 'Legitimacy in the International System', 82 *American Journal of International Law* 705.

Franck, Thomas (1990), *The Power of Legitimacy Among Nations.*

Franck, Thomas (1995), *Fairness in International Law and Institutions.*

Freeman, Michael D.A. (1994), *Lloyd's Introduction to Jurisprudence.*

Freestone, David (1994), 'The Road from Rio: International Environmental Law after the Earth Summit', 6 *Journal of Environmental Law* 193.

Freestone, David and Ellen Hey (eds) (1995), *The Precautionary Principle and International Law: the Challenge of Implementation.*

Frey, Barbara (1977), 'The Legal and Ethical Responsibilities of Transnational Corporations in the Protection of International Human Rights', 6 *Minnesota Journal of Global Trade* 153.

Friedman, Milton (1962), *Capitalism and Freedom.*

Gallarotti, Giulio (1991), 'The Limits of International Organization: Systematic Failure in the Management of International Relations', 45(2) *International Organization* 183.

Germain, Randall (ed.) (2000), *Globalization and Its Critics: Perspectives from Political Economy.*

Giddens, Anthony (1979), *Central Problems in Social Theory: Action, Structure, and Contradiction in Social Analysis.*

Gilhuis, Piet (2002), 'Consequences of the Introduction of Environmental Law Principles in National Law' in Maurice Sheridan and Luc Lavrysen (eds), *Environmental Law Principles in Practice* (2002) 45.

Gillespie, Alexander (2000), *International Environmental Law, Policy and Ethics.*

Gillroy, John Martin (2006), 'Adjudication Norms, Dispute Settlement Regimes and International Tribunals: The Status of "Environmental Sustainability" in International Jurisprudence', 42 *Stanford Journal of International Law* 1.

Ginsburg, Tom (2005), 'Bounded Discretion in International Judicial Lawmaking', 45 *Virginia Journal of International Law* 631.

Gjølberg, Maria and Audun Ruud (2006), 'The UN Global Compact – A Contribution to Sustainable Development?', (Working Paper No 1/05, Centre for Development and Environment, University of Oslo.

Goldman, Emily and Leslie Eliason (eds) (2003), *The Diffusion of Military Technology and Ideas*, p.1.

Goldsmith, Jack and Eric Posner (2005), *The Limits of International Law.*

Goldstein, Judith, Miles Kahler, Robert Keohane and Anne-Marie Slaughter (2000), 'Introduction: Legalization and World Politics', 54(3) *International Organization* 385.

Goldstein, Judith, Miles Kahler, Robert Keohane and Anne-Marie Slaughter (eds) (2001), *Legalization and World Politics.*

Gottlieb, Gidon (1968), *The Logic of Choice: An Investigation of the Concepts of Rule and Rationality.*

Greeves, Lucy and David Ladipo (2004), *Added Values? Measuring the 'Value Relevance' of Sustainability Reporting.*

Grieco, Joseph (1988), 'Anarchy and the Limits of Cooperation: A Realist Critique of the Newest Liberal Institutionalism', 42(3) *International Organization* 485.

Grigorescu, Alexandru (2002), 'European Institutions and Unsuccessful Norm Transmission: The Case of Transparency', 39(4) *International Politics* 467.

Guzzini, Stefano (1993), 'Structural Power: The Limits of Neorealist Power Analysis', 47(3) *International Organization* 443.

Guzzini, Stefano (2000), 'The Use and Misuse of Power Analysis in International Theory' in Ronen Palan (ed.), *Global Political Economy: Contemporary Theories*, p.53.

Guzzini, Stefano and Anna Leander (eds) (2005), *Constructivism and International Relations. Alexander Wendt and His Critics.*

Haas, Ernst (1980), 'Why Collaborate?: Issue-Linkage and International Regimes', 32(3) *World Politics* 357.

Haas, Peter (2002), 'UN Conferences and Constructivist Governance of the Environment', 8 *Global Governance* 73.
Habermas, Jurgen (1981), *Theorie des Kommunikativen Handelns: Vols 1 & 2.*
Habermas, Jurgen (1984), *The Theory of Communicative Action: Vol 2.*
Habermas, Jurgen (1996), *Between Facts and Norms.*
Hafner, Gerhard (2003), 'General Principles of Sustainable Development: From Soft Law to Hard Law' in Fitzmaurice, Malgosia and Szuniewicz, Milena (eds), *Exploitation of Natural Resources in the 21st Century*, p. 53.
Handelman, Don (1995), 'Commentary on Heyman', 36 *Current Anthropology* 280.
Handl, Günther (2007), 'Transboundary Impacts' in Daniel Bodansky , Jutta Brunnée and Ellen Hey (eds) *The Oxford Handbook of International Environmental Law* 531.
Hanqin, Xue (2003), *Transboundary Damage in International Law.*
Harvey, Mark (2000), 'Developing the Agent: Towards a Constructivist Theory of the Identity', Paper presented at the British International Studies Association Conference, Edinburgh.
Held, David and Mathias Koenig-Archibugi (eds) (2005), *Global Governance and Public Accountability.*
Held, David and Anthony McGrew (eds) (2002), *Governing Globalization: Power, Authority and Global Governance.*
Held, David, Anthony McGrew, David Goldblatt, and Jonathan Perraton (1999), *Global Transformations: Politics, Economics and Culture.*
Henkin, Louis (1989), 'International Law: Politics, Values and Functions', 216 *Recueil des Cours* 67.
Herczegh, Geza (1991–1992), *General Principle of Law and the International Legal Order.*
Ellen Hey, 'The Precautionary Concept in Environmental Policy and Law: Institutionalizing Caution' (1991–1992) 4 *Georgetown International Environmental Law Review* 303.
Herczegh, Geza (2000), *Reflections on an International Environmental Court.*
Higgins, Rosalyn (1999), 'Natural Resources in the Case Law of the International Court' in Alan Boyle and David Freestone (eds) (1999), *International Law and Sustainable Development: Past Achievements and Future Challenges*, p.101.
Hjerner, Lars, Ramberg Jan, Ove Bring and Said Mahmoudi (1990), *Festskrift Till Lars Hjerner: Studies in International Law.*
Hohmann, Harold (1994), *Precautionary Legal Duties and Principles of Modern International Environmental Law.*
Hollis, Martin and Steve Smith (1990), *Explaining and Understanding International Relations.*

Hooghe, Liesbet and Gary Marks (2001), *Multi-Level Governance and European Integration*.

Horbach, Nathalie and Pieter Bekker (2003), 'State Responsibility for Injurious Transboundary Activity in Retrospect', 50 *Netherlands International Law Review* 327.

Hunt, Alan (1993), *Explorations in Law and Society: Toward a Constitutive Theory of Law*.

Hunter, David, James Salzman and Durwood Zaelke (1998), *International Environmental Law and Policy*.

Hurrell, Andrew (1998), 'Society and Anarchy in International Relations' in Barbara Allen Roberson (ed.), *International Society and the Development of International Relations Theory*, p.17.

Hurrell, Andrew (2002), 'Foreword to the Third Edition: The Anarchical Society 25 Years On' in Hedley, Bull, *The Anarchical Society: a Study of Order in World Politics*.

Hurrell, Andrew (2005), 'Norms and Ethics in International Relations' in Walter Carlsnaes, Thomas Risse-Kappen, and Beth Simmons (eds), *Handbook of International Relations* (2005).

Ikenberry, John (1990), 'The International Spread of Privatization Policies: Inducements, Learning, and Policy Bandwagoning' in Ezra Suleiman and John Waterbury (eds), *The Political Economy of Public Sector Reform and Privatization*, p.99.

Jackson, Bernard (1988), *Law, Fact and Narrative Coherence*.

Jackson, Robert (2000), *The Global Covenant: Human Conduct in a World of States*.

Jacobsson, Kerstin (2004), 'Between the Deliberation and Discipline: Soft Governance in EU Employment Policy' in Ulrika Mörth (ed.), *Soft Law in Governance and Regulation: An Interdisciplinary Analysis*, p.81.

Jacobsson, Kerstin (2004), 'Soft Regulation and the Subtle Transformation of States: The Case of the EU Employment Policy', 14(4) *Journal of European Social Policy* 355.

Jalet, Frances (1963), 'The Quest for the General Principles of Law Recognized by Civilized Nations – A Study', 10 *UCLA Law Review* 1041.

Jennings, Robert (1996), 'The Judiciary, International and National, and the Development of International Law', 45 *International and Comparative Law Quarterly* 1.

Jepperson, Ronald, Alexander Wendt, and Peter Katzenstein (1996), 'Norms, Identity, and Culture in National Security' in Peter Katzenstein (ed.), *The Culture of National Security. Norms and Identity in World Politics*, p.33.

John, Fleur (1994), 'The Invisibility of the Transnational Corporation:

An Analysis of International Law and Legal Theory', 19 *Melbourne University Law Review* 893.

Johnston, Alastair (2001), 'Treating International Institutions as Social Environments', 45(4) *International Studies Quarterly* 487.

Johnston, Alastair (2005), 'Conclusions and Extensions: Toward Mid-Range Theorizing and Beyond Europe', 59 *International Organization* 1013.

Johnston, Douglas (1999), 'Fishery Diplomacy and Science and the Judicial Function', 10 *Yearbook of International Environmental Law* 33.

Johnston, Douglas (2006), 'The challenge of international ocean governance: institutional, ethical and conceptual dilemmas' in Donald Rothwell and David VanderZwagg (eds), *Towards Principled Oceans Governance: Australian and Canadian approaches and challenges*, p.349.

Johnstone, Ian (1991), 'Treaty Interpretation: The Authority of Interpretive Communities', 12 *Michigan Journal of International Law* 371.

Jörgens, Helge (2004), 'Governance by Diffusion: Implementing Global Norms Through Cross-national Imitation and Learning' in William Lafferty, *Governance for Sustainable Development: The Challenges of Adapting Form to Function*, p.246.

Kahneman, Daniel, Paul Slovic and Amos Tversky (1982), *Judgment Under Uncertainty: Heuristics and Biases*.

Katzenstein, Peter (ed.), *The Culture of National Security: Norms and Identity in World Politics*.

Katzenstein, Peter, 'Introduction: Alternative Perspectives on National Security' in, Peter Katzenstein (ed.), *The Culture of National Security: Norms and Identity in World Politics*, p.1.

Kaya, Ibrahim (2003), *Equitable Utilization: The Law of the Non-Navigational Uses of International Watercourses*.

Keck, Margaret and Kathryn Sikkink (1998), *Activists Beyond Borders: Advocacy Networks in International Politics*.

Keck, Margaret and Kathryn Sikkink (1999), 'Transnational advocacy networks in international and regional politics', 51 *International Social Science Journal* 89.

Kell, Georg (2005), 'The Global Compact, Selected Experiences and Reflections', 59 *Journal of Business Ethics* 69.

Kell, Georg and David Levin (2003), 'The Global Compact Network: An Historic Experiment in Learning and Action', 108(2) *Business and Society Review* 151.

Kell, Georg and John Ruggie (1999), 'Global Markets and Social Legitimacy: the Case for the "Global Compact"', 8(3) *Transnational Corporations* 101.

Kennedy, Duncan (1997), *A Critique of Adjudication (fin de siecle)*.

Keohane, Nathaniel, Richard Revesz and Robert Stavins (1988), 'The Choice of Regulatory Instruments in Environmental Policy', 22 *Harvard Environmental Law Review* 313.

Keohane, Robert (1984), *After Hegemony: Cooperation and Discord in the World Political Economy*.

Keohane, Robert (1989), *International Institutions and State Power*.

Keohane, Robert, Andrew Moravcsik and Anne-Marie Slaughter (2000), 'Legalized Dispute Resolution: Interstate and Transnational', 54(3) *International Organization* 457.

Kern, Kristine, Helge Jörgens and Martin Jänicke (2001), 'The Diffusion of Environmental Policy Innovations: A Contribution to the Globalisation of Environmental Policy', WZB Discussion Paper FS II 01-203, Berlin, Social Science Research Centre.

Kihlström, Anita and Joakim Israel (2002), 'Communicative or Strategic Action – an Examination of Fundamental Issues in the Theory of Communicative Action', 11(3) *International Journal of Social Welfare* 210.

Kindt, John (1986), *Marine Pollution and the Law of the Sea*.

King, Betty (2001), 'The UN Global Compact: Responsibility for Human Rights, Labor Relations, and the Environment in Developing Nations', 34 *Cornell International Law Journal* 481.

Kingsbury, Benedict (1998), '"Indigenous Peoples" in International Law: A Constructivist Approach to the Asian Controversy', 92(3) *The American Journal of International Law* 414.

Kingsbury, Benedict (1998), 'The Concept of Compliance as a Function of Competing Conceptions of International Law', 19 *Michigan Journal of International Law* 345.

Kingsbury, Benedict (1999), 'Is the Proliferation of International Courts and Tribunals a Systemic Problem', 31(4) *New York University Journal of International Law and Politics* 679.

Kiss, Alexandre and Dinah Shelton (2004), *International Environmental Law*.

Kiss, Alexandre and Dinah Shelton (2007), *Guide to International Environmental Law*.

Klabbers, Jan (1996), 'The Redundancy of Soft Law', 65(2) *Nordic Journal of International Law* 167.

Klabbers, Jan (1998), 'The Undesirability of Soft Law', 67(4) *Nordic Journal of International Law* 381.

Klabbers, Jan (2005), 'Reflections on Soft International Law in a Privatized World', 16 *Finnish Yearbook of International Law* 313.

Klein, Natalie (2005), *Dispute Settlement in the UN Convention on the Law of the Sea*.

Klotz, Audie (2002), 'Transnational Activism and Global Transformations: The Anti-Apartheid and Abolitionist Experiences', 8(1) *European Journal of International Relations* 49.

Knill, Christoph and Dirk Lehmkuhl (2002), 'Private Actors and the State: Internationalization and Changing Patterns of Governance', 15 *Governance* 41.

Koskenniemi, Martti (1991), 'Peaceful Settlement of Environmental Disputes', 60 *Nordic Journal of International Law* 73.

Koskenniemi, Martti (2000), 'General principles: reflexions on constructivist thinking in international law' in Koskenniemi, Martti (ed), *Sources of International Law*, p.359.

Koskenniemi, Martti (2005), 'International Legislation Today: Limits and Possibilities', 23 *Wisconsin International Law Journal* 61.

Koskenniemi, Martti (2006), *From Apology to Utopia: The Structure of International Legal Argument*.

Koskenniemi, Martti (2007), 'The Fate of Public international Law: Between Technique and Politics', 70 *Modern Law Review* 1.

Kratochwil, Friedrich (1991), *Rules, Norms, and Decisions: On the Conditions of Practical and Legal Reasoning in International Relations and Domestic Affairs*.

Kratochwil, Friedrich (2000), 'How Do Norms Matter?' in Michael Byers (ed.), *The Role of Law in International Politics: Essays in International Relations and International Law*, p.34.

Kratochwil, Friedrich and John Ruggie (1986), 'International Organization: A State of the Art on an Art of the State', 40 *International Organization* 768.

Krebs, Ronald and Patrick Jackson (2007), 'Twisting Tongues and Twisting Arms: The Power of Political Rhetoric', 13(1) *European Journal of International Relations* 35.

Kritsiotis, Dino (2002), 'Imagining the International Community', 13(4) *European Journal of International Law* 961.

Kritsiotis, Dino (2004), 'When States Use Armed Force' in Christian Reus-Smit (ed.), *The Politics of International Law*, p.45.

Lang, Winfried (1999), 'UN-Principles and International Environmental Law', 3 *Max Planck United Nations Yearbook* 157.

Lave, Jean and Etienne Wenger (1991), *Situated learning: Legitimate peripheral participation*.

Leigh, Kathy (1992), 'Liability for Damage to the Global Commons', 14 *Australian Year Book of International Law* 129.

Leipziger, Deborah (2003), *The Corporate Responsibility Code Book*.

Leisinger, Klaus (2003), 'Opportunities and Risks of the United Nations Global Compact', 11 *Journal of Corporate Citizenship* 113.

Lenschow, Andrea (ed.) (2000), *Environmental Policy Integration: Greening Sectoral Policies in Europe*.

Lesser, Eric, Michael Fontaine and Jason Slusher (eds) (2000), *Knowledge and Communities*.

Levy, Jack (1984), 'Learning and Foreign Policy: Sweeping a Conceptual Minefield', 48(2) *International Organization* 279.

Lillich, Richard (ed.) (1995), *The United Nations Compensation Commission*.

Linklater, Andrew (2005), 'Dialogic Politics and the Civilising Process', 31(1) *Review of International Studies* 141.

Linklater, Andrew and Hidemi Suganami (2006), *The English School of International Relations: A Contemporary Reassessment*.

Lose, Lars (2001), 'Communicative Action and the World of Diplomacy' in Fierke, Karin and Knud Jørgensen (eds), *Constructing International Relations: the next generation*, p.179

Louka, Elli (2006), *International Environmental Law: Fairness, Effectiveness, and World Order*.

Lowe, Vaughan (1999), 'Sustainable Development and Unsustainable Arguments' in Alan Boyle and David Freestone (eds), *International Law and Sustainable Development: Past Achievements and Future Challenges*, p.19.

Lowe, Vaughan (2000), 'The Politics of Law-Making: Are the Method and Character of Norm Creation Changing?' in Michael Byers (ed.), *The Role of Law in International Politics: Essays in International Relations and International Law*, p.58.

MacCormick, Neil (1978), *Legal Reasoning and Legal Theory*.

Macrory, Richard (ed.), *Principles of European Environmental Law*.

Maes, Frank (2002), 'Environmental Law Principles, their Nature and the Law of the Sea: A Challenge for Legislators' in Maurice Sheridan and Luc Lavrysen (eds), *Environmental Law Principles in Practice*, p.59.

Magraw, Daniel and Lisa Hawke (2007), 'Sustainable Development' in Daniel Bodansky, Jutta Brunnée, and Ellen Hey (eds), *The Oxford Handbook of International Environmental Law*, p.613.

Makarczyk, Jerzy (ed.) (1996), *Theory of International Law at the Threshold of the 21st Century: Essays in Honour of Krzysztof Skubiszewski*.

Mansfield, Bill (2001), 'Letter to the Editor: The Southern Bluefin Tuna Arbitration: Comments on Professor Barbara Kwiatkowska's Article', 16(2) *International Journal of Marine and Coastal Law* 361.

March, James and Johan Olsen (1998), 'The Institutional Dynamics of International Orders', 52 *International Organization* 943.

March, James and Johan Olsen (1989), *Rediscovering Institutions: The Organizational Basis of Politics*.

Marcus, David (2006), 'The Normative Development of Socioeconomic Rights Through Supranational Adjudication', 42 *Stanford Journal of International Law* 53.

Marong, Alhaji (2003), 'From Rio to Johannesburg: Reflections on the Role of International Legal Norms in Sustainable Development', 16(1) *Georgetown International Environmental Law Review* 21.

Martin, Steve (ed.) (1984), *The Construction of Europe – Essays in honour of Emile Noel.*

McCaffrey, Stephen (2007), *The Law of International Watercourses.*

McCaffrey, Stephen and Robert Rosenstock (1996), 'The International Law Commission's Draft Articles on International Watercourses: An Overview and Commentary', 5(2) *Review of European Community and International Environmental Law* 89.

McCaffrey, Stephen and Mpazi Sinjela (1998), 'The 1997 United Nations Convention on International Watercourses' 92 *American Journal of International Law* 97.

McCloskey, Michael (1998), 'The Emperor Has No Clothes: The Conundrum of Sustainable Development', 9 *Duke Environmental Law and Policy Forum* 153.

McDonald, R. St J. and B. Hough (1977), 'The Nuclear Tests Case Revisited', 20 *German Yearbook of International Law* 337.

McDonald, Jan (2006) 'Tr(e)adding cautiously: precaution in WTO decision making' in Elizabeth Fisher, Judith Jones, Rene von Schomberg (eds) *Implementing the Precautionary Principle: Perspective and Prospects*, p.160.

McDougal, Myres and William Burke (1961), *The Public Order of the Oceans: A Contemporary International Law of the Sea.*

McIntyre, Owen (2004), 'The Emergence of an "Ecosystem Approach" to the Protection of International Watercourses under International Law', 13 *Review of European Community & International Environmental Law* 1.

McIntyre, Owen and Thomas Mosedale (1997), 'The Precautionary Principle as a Norm of Customary International Law', 9 *Journal of Environmental Law* 221.

Mearsheimer, John (1994–1995), 'The False Promise of International Institutions', 19 *International Security* 5.

Merrills, John (2006), *International Dispute Settlement.*

Meyer, William and Boyka Stefanova (2001), 'Human Rights, the UN Global Compact, and Global Governance', 34 *Cornell International Law Journal* 501.

Miller, Clark (2007), 'Democratization, International Knowledge Institutions, and Global Governance', 20 *Governance* 325.

Miller, Clark and Paul Edwards (eds) (2001), *Changing the Atmosphere: Expert Knowledge and Environmental Governance*.

Miller, Nathan (2002), 'An International Jurisprudence? The Operation of 'Precedent' Across International Tribunals', 15 *Leiden Journal of International law* 483.

Milliken, Jennifer (1999), 'The Study of Discourse in International Relations: A Critique of Research and Methods', 5 *European Journal of International Relations* 225.

Mitchell, Ronald (2002), 'International Environment' in Walter Carlsnaes, Thomas Risse and Beth A. Simmons (eds), *The Handbook of International Relations*, p. 500.

Mitzen, Jennifer (2005), 'Reading Habermas in Anarchy: Multilateral Diplomacy and Global Public Spheres', 99(3) *American Political Science Review* 401.

Moore, Michael (1996–97), 'Legal Principles Revisited', 82 *Iowa Law Review* 867.

Moravcsik, Andrew (2001), 'Bringing Constructivist Integration Theory Out Of The Clouds: Has It Landed Yet?', 2(2) *European Union Politics* 226.

Morgera, Elisa (2006), 'An Environmental Outlook on the OECD Guidelines for Multinational Enterprises: Comparative Advantage, Legitimacy, and Outstanding Questions in the Lead Up to the 2006 Review', 18 *Georgetown International Environmental Law Review* 751.

Morgera, Elisa (2006), 'The UN and Corporate Environmental Responsibility: Between International Regulation and Partnerships', 15 *Review of European Community and International Environmental Law* 93.

Mörth, Ulrika (ed.) (2004), *Soft Law in Governance and Regulation: An Interdisciplinary Analysis*.

Mörth, Ulrika (ed.) (2004), 'Conclusions' in Mörth, Ulrika (ed.), *Soft Law in Governance and Regulation: An Interdisciplinary Analysis*, p.191.

Mosler, Hermann (1980), *The International Society as a Legal Community*.

Muchlinski, Peter (1995), *Multinational Enterprises and the Law*.

Muchlinski, Peter (1997), '"Global Bukowina" Examined: Viewing the Multinational Enterprise as a Transnational Law-Making Community' in Gunther Teubner (ed.), *Global Law Without a State*, p.79.

Muchlinski, Peter (2000), 'Attempts to Extend the Accountability of transnational Corporations: The Role of UNCTAD' in Menno Kamminga and Saman Zia-Zarifi (eds), *Liability of Multinational Corporations Under International Law*, p.97.

Müller, Harald (2004), 'Arguing, Bargaining and All That: Communicative

Action, Rationalist Theory and the Logic of Appropriateness in International Relations', 10(3) *European Journal of International Relations* 395.

Müller, Harald (2001), International Relations as Communicative Action' in Karin Fierke and Knud Jørgensen (eds), *Constructing International Relations. The Next Generation*, p.60.

Murphy, Sean (2005), 'Taking Multinational Corporate Codes of Conduct to the Next Level', 43 *Columbia Journal of Transnational Law* 389.

Nadelmann, Ethan (1990), 'Global Prohibition Regimes: The Evolution of Norms in International Society', 44 *International Organization* 479.

Nollkaemper, André (1995), '"What you risk reveals what you value", and Other Dilemmas Encountered in the Legal Assaults on Risks' in David Freestone and Ellen Hey (eds), *The Precautionary Principle and International Law*, 73.

Nollkaemper, André (2000), 'Three Conceptions of the Integration Principle in International Environmental Law' in Andrea Lenschow (ed.), *Environmental Policy Integration: Greening Sectoral Policies in Europe*, p.22.

Nowotny, Helga, Peter Scott and Michael Gibbons, *Re-Thinking Science: Knowledge and the Public in an Age of Uncertainty*.

Nye, Joseph (2004), *Soft Power: The Means to Success in World Politics*.

Nye, Joseph and John Donahue (2000), *Governance in a Globalizing World*.

O'Riordan, Timothy (ed.) (1997), *Ecotaxation*.

Onuf, Nicholas (1985), 'Do Rules Say What They Do? From Ordinary Language to International Law', 26 *Harvard International Law Journal* 385.

Onuf, Nicholas (1989), *World of Our Making: Rules and Rule in Social Theory and International Relations*.

Orbell, John, Alphons Van de Kragt and Robyn Dawes (1988), 'Explaining discussion-induced cooperation', 34 *Journal of Personality and Social Psychology* 811.

Orbell, John, Robyn Dawes and Alphons Van de Kragt (1990), 'The limits of multilateral promising', 100(3) *Ethics* 616.

Palan, Ronen (ed.) (2000), *Global Political Economy: Contemporary Theories*.

Palan, Ronen (ed.) (2000), 'Recasting Political Authority: Globalization and the State' in Randall Germain (ed.), *Globalization and Its Critics: Perspectives from Political Economy*, p.139.

Pallemaerts, Marc (1992), 'International Environmental Law from Stockholm to Rio: Back to the Future', 1 *Review of European Community and International Environmental Law* 254.

Pallemaerts, Marc (1993), 'International Environmental Law From Stockholm to Rio: Back to the Future?' in Philippe Sands (ed.), *Greening International Law*, p.1.

Panjabi, Ranee (1992–1993), 'From Stockholm To Rio: A Comparison of the Declaratory Principles of International Environmental Law', 21 *Denver Journal of International Law and Policy* 215.

Paradell-Trius, Lluís (2000), 'Principles of International Environmental Law: an Overview', 9(2) *Review of European Community & International Environmental Law* 93.

Park, Susan (2006), 'Theorizing Norm Diffusion Within International Organizations', 43 *International Politics* 342.

Parker, Christine, Colin Scott, Nicola Lacey and John Braithwaite (eds) (2004), *Regulating Law*.

Pattberg, Philipp (2006), 'The Influence of Global Business Regulation: Beyond Good Corporate Conduct', 111 *Business and Society Review* 241.

Paulus, Andreas L. (2005), 'Jus Cogens in a Time of Hegemony and Fragmentation – An Attempt at a Re-appraisal', 74 (3–4) *Nordic Journal of International Law* 297.

Payne, Rodger (2001), 'Persuasion, Frames and Norm Construction', 7 *European Journal of International Relations* 37.

Peczenik, Aleksander (1971), 'Principles of Law: The Search for Legal Theory', 2 *Rechtstheorie* 17.

Peel, Jacqueline (2004), 'Precaution – A Matter of Principle, Approach or Process?', 5 *Melbourne Journal of International Law* 483.

Penalver, Eduardo Moises (2000), 'The Persistent Problem of Obligation in International Law', 36 *Stanford Journal of International Law* 271.

Piaget, Jean (1932), *The Moral Judgment of the Child.*

Picciotto, Sol (1999), 'Introduction: What Rules for the World Economy?' in Sol Picciotto and Ruth Mayne (eds), *Regulating International Business: Beyond Liberalization*, p.1

Picciotto, Sol and Ruth Mayne (eds) (1999), *Regulating International Business: Beyond Liberalization.*

Pinto, Moragodage (1996), '"Common Heritage of Mankind": From Metaphor to Myth, and the Consequences of Constructive Ambiguity' in Jerzy Makarczyk (ed.), *Theory of International Law at the Threshold of the 21st Century: Essays in Honour of Krzysztof Skubiszewski*, p.249.

Podehl, Jorg and Donald Rothwell (1992), 'New Zealand and the Convention on the Regulation of Antarctic Mineral Resource Activities (CRAMRA): An Unhappy Divorce?', 22 *Victoria University of Wellington Law Review* 23.

Porras, Ileana (1993), 'The Rio Declaration: A New Basis for International

Cooperation' in Philippe Sands (ed.), *Greening International Law* (1993), p.20.

Postiglione, Amedeo (1996), *The Global Environmental Crisis: The Need for an International Court of the Environment*.

Powell, Robert (1991), 'Absolute and Relative Gains in International Relations Theory', 85 *International Organization* 1303.

Preiss, Erika (1999), 'The International Obligation to Conduct an Environmental Impact Assessment: The ICJ Case Concerning the Gabčíkovo-Nagymaros Project', 7 *New York University Environmental Law Journal* 307.

Price, Richard and Nina Tannenwald (1996), 'Norms and Deterrence: The Nuclear and Chemical Weapons Taboos' in Peter Katzenstein (ed.), *The Culture of National Security: Norms and Identity in World Politics*, p.114.

Price, Richard and Nina Tannenwald (1998), 'Reversing the Gun Sights: Transnational Civil Society Targets Land Mines', 52 *International Organization* 613.

Rajamani, Lavanya (2000), 'The Principle of Common but Differentiated Responsibility and the Balance of Commitments under the Climate Regime', 9 *Review of European Community and International Environmental Law* 120.

Rao, Pemmaraju Sreenivasa (1997), 'Environment as a Common Heritage of Mankind: a Policy Perspective' in *International Law on the Eve of the Twenty-First Century: Views from the International Law Commission: réflexions de codificateurs*, p.201.

Rao, Pemmaraju Sreenivasa (2004), 'International Liability for Transboundary Harm', 34(6) *Environmental Policy and Law* 224.

Ratner, Steven (2000), 'Does International Law Matter in Preventing Ethnic Conflict?', 32 *New York University Journal of International Law and Politics* 591.

Raz, Joseph (1972), 'Legal Principles and the Limits of the Law', 81 *Yale Law Journal* 823.

Raz, Joseph (1999), *Practical Reason and Norms*.

Rest, Alfred (1998), 'The Indispensability of an International Environmental Court', 7 *Review of European Community and International Environmental Law* 63.

Reus-Smit, Christian (1999), *The Moral Purpose of the State*.

Reus-Smit, Christian (2001), 'Human Rights and the Social Construction of Sovereignty', 27 *Review of International Studies* 519.

Reus-Smit, Christian (2001), 'Politics and International Legal Obligation', 9(4) *European Journal of International Relations* 591.

Reus-Smit, Christian (ed.) (2004), *The Politics of International Law*.

Reus-Smit, Christian (2004), 'Introduction' in Christian Reus-Smit (ed.), *The Politics of International Law*, p.1.

Reus-Smit, Christian (2004), 'The Politics of International Law' in Christian Reus-Smit (ed.), *The Politics of International Law*, p.14.

Ringmar, Erik (1997), 'Alexander Wendt – a Social Scientist Struggling with History' in Iver Neumann and Ole Waever (eds), *The Future of International Relations: Masters, in the Making*, p.269.

Risse, Thomas (1997), '"Let's Talk!" Insights from the German Debate on Communicative Behavior and International Relations', Paper presented to the Annual Convention of the American Political Science Association, Washington DC, 27–31 August 1997.

Risse, Thomas (2000), '"Let's Argue!": Communicative Action in World Politics', 54(1) *International Organization* 1.

Risse, Thomas (2004), 'Global Governance and Communicative Action', 39(2) *Government and Opposition* 288.

Risse, Thomas (2005), 'Global Governance and Communicative Action' in David Held and Mathias Koenig-Archibugi (eds), *Global Governance and Public Accountability*, p.164.

Risse, Thomas, Stephen Ropp and Kathryn Sikkink (eds) (1999), *The Power of Human Rights: International Norms and Domestic Change.*

Rogers, Everett (1995), *Diffusion of Innovations.*

Roman, Ediberto (1998), 'Reconstructing Self-Determination: The Role of Critical Theory in the Positivist International Law Paradigm', 53 *University of Miami Law Review* 943.

Romano, Cesare (2000), *The Peaceful Settlement of International Environmental Disputes.*

Romano, Cesare (2007), 'International Dispute Settlement' in Daniel Bodansky, Jutta Brunnée, and Ellen Hey (eds), *The Oxford Handbook of International Environmental Law*, p.1036.

Rosenau, James (2006), *The Study of World Politics: Globalization and Governance (Vol. 2).*

Rothwell, Donald and David VanderZwaag (eds) (2006), *Towards Principled Oceans Governance: Australian and Canadian approaches and challenges.*

Ruggie, John (1992), 'Multilateralism: the Anatomy of an Institution' 46(3) *International Organization* 561.

Ruggie, John (1993), 'Multilateralism: The Anatomy of an Institution', in John Ruggie (ed.), *Multilateralism Matters: the Theory and Praxis of an Institutional Form*, p.3.

Ruggie, John (1998), 'What Makes the World Hang Together? Neo-utilitarianism and the Social Constructivist Challenge', 52 *International Organization* 855.

Ruggie John (1998), *Constructing the World Polity: Essays on International Institutionalization.*

Ruggie John (2002), 'The Theory and Practice of Learning Networks: Corporate Social Responsibility and the Global Compact', 5 *Journal of Corporate Citizenship* 27.

Ruggie John (2004), 'Reconstituting the Global Public Domain – Issues, Actors, and Practices' 10(4) *European Journal of International Relations* 499.

Ruiz, J.J. (1977), 'Mootness in International Adjudication: The Nuclear Tests Cases', 20 *German Yearbook of International Law* 358.

Sagoff, Mark (1988), *The Economy of the Earth: Philosophy, Law and the Environment.*

Sagoff, Mark (2004), *Price, Principle, and the Environment.*

Sandholtz, Wayne and Alec Stone Sweet (2004), 'Law, Politics, and International Governance' in Christian Reus-Smit (ed.), *The Politics of International Law* 238.

Sands, Philippe (ed.) (1993), *Greening International Law.*

Sands, Philippe (ed.) (1995), 'International Law in the Field of Sustainable Development: Emerging Legal Principles' in Winfried Lang (ed.), *Sustainable Development and International Law*, p.53.

Sands, Philippe (ed.) (1998), 'Pleadings and the Pursuit of International Law: Nuclear Tests II (*New Zealand v France*)' in Antony Anghie and Garry Sturgess (eds), *Legal Visions of the 21st Century: Essays in Honour of Judge Christopher Weeramantry*, p.601.

Sands, Philippe (ed.) (1999), *Transnational Environmental Law: Lessons in Global Change.*

Sands, Philippe (ed.) (1999), 'International Courts and the Application of the Concept of "Sustainable Development"', *Max Planck Yearbook of United Nations Law*, p.389.

Sands, Philippe (ed.) (2003), *Principles of International Environmental Law.*

Schelling, Thomas (1966), *Arms and Influence.*

Schepel, Harm (2005), *The Constitution of Private Governance: Product Standards in the Regulation of Integrating Markets.*

Schoppa, Leonard (1999), 'The Social Context in Coercive International Bargaining', 53(2) *International Organization* 307.

Schrijver, Nico (1988), 'Permanent sovereignty over national resources versus the common heritage of mankind: complementary or contradictory principles of international economic law?' in Paul De Waart, Paul Peters and Erik Denters (eds), *International Law and Development*, p.87.

Schulte, Constanze (2004), *Compliance with Decisions of the International Court of Justice.*

Schwabach, Aaron (2006), *International Environmental Disputes: A Reference Handbook*.

Schwartz, Thomas (1978), 'Obligations to Posterity' in Richard Sikora and Brian Barry (eds), *Obligations to Future Generations*.

Schwarzenberger, Georg (1987), *The Fundamental Principles of International Law*.

Schwebel, Stephen 'Ad Hoc Chambers of the International Court of Justice', 81 *American Journal of International Law* 831.

Searle, John (1995), *The Construction of Social Reality*.

Shaffer, Gregory (2005), 'Power, governance, and the WTO: a comparative institutional approach' in Michael Barnett and Raymond Duvall (eds), *Power in Global Governance* (2005) 130.

Shahabuddeen, Mohamed (1996), *Precedent in the World Court*.

Sheridan, Maurice and Luc Lavrysen (eds) (2002), *Environmental Law Principles in Practice*.

Shore, Cris and Susan Wright (1997), 'Policy: A New Field of Anthropology' in Cris Shore and Susan Wright (eds), *Anthropology of Social Policy: Critical Perspectives on Governance and Power*, p.3.

Sikkink, Kathryn (1986), 'Codes of Conduct for Transnational Corporations: the Case of the WHO/UNICEF Code', 40 *International Organization* 815.

Sikora, Richard and Brian Barry (eds) (1978), *Obligations to Future Generations*.

Simma, Bruno (2002), 'The Work of the International Law Commission at Its Fifty-Third Session (2001)', 71 *Nordic Journal of International Law* 123.

Simma, Bruno and Andreas Paulus (1998), 'The "International Community": Facing the Challenge of Globalization', 9 *European Journal of International Law* 266.

Singh, Nagendra (1987), 'Foreword' in World Commission on Environment and Development, R.D. Munro and J. G. Lammers, *Environmental Protection and Sustainable Development: Legal Principles and Recommendations*, p.1.

Singh, Nagendra (1988), 'Sustainable development as a principle of international law' in Paul De Waart, Paul Peters and Erik Denters (eds), *International Law and Development*, p.1.

Snyder, Francis (1994), 'Soft Law and Institutional Practice in the European Community' in, Steve Martin (ed.), *The Construction of Europe – Essays in Honour of Emile Noel*, p.197.

Sohn, Louis (1973), 'The Stockholm Declaration on the Human Environment', 14 *Harvard International Law Journal* 423.

Solomon, Scott (2006), 'Power and the Politics of Global Governance', 8 *International Studies Review* 327.

Steinberg, Richard and Jonathan Zasloff (2006), 'Power and International Law', 100(1) *The American Journal of International Law* 64.
Stephens, Tim (2004), 'The Limits of International Adjudication in International Environmental Law: Another Perspective on the Southern Bluefin Tuna Case', 19(2) *The International Journal of Marine and Coastal Law* 177.
Stephens, Tim (2005), *The Role of International Courts and Tribunals in International Environmental Law*, D Phil Thesis, University of Sydney.
Stephens, Tim (2006), 'Multiple International Courts and the "Fragmentation" of International Environmental Law', 25 *Australian Yearbook of International Law* 227.
Stephens, Tim (2008), *International Courts and Environmental Protection.*
Stewart, Richard (2001), 'A New Generation of Environmental Regulation?', 29 *Capital University Law Review* 21.
Stewart, Richard (2007), 'Instrument Choice' in Daniel Bodansky, Jutta Brunnée and Ellen Hey (eds), *The Oxford Handbook of International Environmental Law*, p.147.
Stone, Christopher (1985), '*Should Trees Have Standing?* Revisited: How Far Will Law and Morals Reach? A Pluralist Perspective', 59 *Southern California Law Review* 1.
Stone, Christopher (1988), 'The Environment in Moral Thought', 56 *Tennessee Law Review* 1.
Stone, Christopher (2004), 'Common but Differentiated Responsibilities in International Law', 98 *American Journal of International Law* 276.
Stone, Christopher (2007), 'Ethics and International Environmental Law' in Daniel Bodansky, Jutta Brunnée and Ellen Hey (eds) *The Oxford Handbook of International Environmental Law*, p.291.
Stone Sweet, Alee (1999), 'Judicialization and the Construction of Governance', 32 *Comparative Political Studies* 2.
Strang, David and John Meyer (1993), 'Institutional conditions for diffusion', 22 *Theory and Society* 487.
Strange, Susan (1996), *The Retreat of the State: The Diffusion of Power in the World Economy.*
Strange, Susan (2003), *States and Markets.*
Sztucki, Jerzy (1990), 'Reflections on International "Soft Law"' in Lars Hjerner, Jan Ramberg, Ove Bring and Said Mahmoudi, *Festskrift Till Lars Hjerner: Studies in International Law*, p.549.
Tannenwald, Nina (1999), 'The Nuclear Taboo: The United States and the Normative Basis of Nuclear Non-Use', 53(3) *International Organization* 433.
Tavis, Lee (2002), 'Corporate Governance and the Global Social Void', 35 *Vanderbilt Journal of Transnational Law* 487.

Taylor, Charles (1989), *Sources of the Self: The Making of the Modern Identity*.

Taylor, Charles (1993), 'To Follow a Rule . . .' in Craig Calhoun, Edward LiPuma and Moishe Postone (eds), *Bourdieu: Critical Perspectives*, p.45

Taylor, Prue (1998), *An Ecological Approach to International Law: Responding to challenges of climate change*.

Tews, Kerstin, Per-Olof Busch and Helge Jorgens (2003), 'The Diffusion of New Environmental Policy Instruments', 42(4) *European Journal of Political Research* 569.

Thirlway, Hugh (1989), 'The Law and Procedure of the International Court of Justice', 60 *British Yearbook of International Law* 1.

Toope, Stephen (2000), 'Emerging Patterns of Governance and International Law' in Michael Byers, (ed.), *The Role of Law in International Politics: Essays in International Relations and International Law*, p.91.

Toope, Stephen (2007), 'Formality and Informality' in Daniel Bodansky, Jutta Brunnée and Ellen Hey (eds), *The Oxford Handbook of International Environmental Law*, 107.

Trachtman, Joel (1999), 'The Domain of WTO Dispute Resolution', 40 *Harvard International Law Journal* 333.

Tromp, D. (1991), 'Foreword' in Office for the London Dumping Convention at the International Maritime Organization, *The London Convention: The First Decade and Beyond*, p.1.

Trouwborst, Arie (2002), *Evolution and Status of the Precautionary Principle in International Law*.

Trouwborst, Arie (2006), *Precautionary Rights and Duties of States*.

Trouwborst, Arie (2007), 'The Precautionary Principle in General International Law: Combating the Babylonian Confusion', 16 *Review of European Community & International Environmental Law* 185.

Tushnet, Mark (1983), 'Following the Rules Laid Down: A Critique of Interpretivism and Neutral Principles', 96 *Harvard Law Review* 781.

Twining, William and David Miers (1999), *How To Do Things With Rules*.

Ulbert, Cornelia and Thomas Risse (2005), 'Deliberately Changing the Discourse: What Does Make Arguing Effective', 40 *Acta Politica* 351.

Ulbert, Cornelia, Thomas Risse and Harald Müller (2004), 'Arguing and Bargaining in Multilateral Negotiations', Paper presented at the Conference on Empirical Approaches to Deliberative Politics, France, 21–22 May.

Ulbert, Cornelia and Simone Wisotzki (2001), '(Pre-)structuring Argumentative Processes: the Impact of Institutional Settings on Banning Worst Forms of Child Labour and Landmines', Paper presented at the

4th Pan European International Relations Conference, University of Kent, Canterbury, 8–10 September.
United Nations Department of Economic and Social Affairs (1974), *The Impact of Multinational Corporations on Development and on International Relations*.
van Asselt, Marjolein and Ellen Vos (2006), 'The Precautionary Principle and the Uncertainty Paradox', 9(4) *Journal of Risk Research* 313.
van Dijk, Pieter (1987), 'Normative Force and Effectiveness of International Norms', 30 *German Yearbook of International Law* 14.
van Gestel, Rob and Guido Evers (2005), 'Communicative legislation: Can we trust certified management systems as tools for the interpretation and enforcement of environmental laws?' in Nicolle Zeegers, Willem Witteveen and Bart van Klink (eds), *Social and Symbolic Effects of Legislation Under the Rule of Law*, p.183.
van Hoecke, Mark (2002), *Law as Communication*.
VanderZwaag, David (1999), 'The Precautionary Principle in Environmental Law and Policy: Elusive Rhetoric and First Embraces', 8 *Journal of Environmental Law and Practice* 355.
Verschuuren, Jonathan (2003), *Principles of Environmental Law: The Ideal of Sustainable Development and the Role of Principles of International, European, and National Environmental Law*.
Wæver, Ole (1998), 'Four Meanings of International Society: A Trans-Atlantic Dialogue' in Barbara Allen Roberson (ed.), *International Society and the Development of International Relations Theory*, p.80.
Wagner, Peter, Carol Hirschon Weiss, Bjorn Wittrock and Hellmut Wollmann (eds) (1991), *Social Sciences and Modern States: National Experiences and Theoretical Crossroads*.
Walsh, Virginia (2004), *Global Institutions and Social Knowledge: Generating Knowledge at the Scripps Institution and the Inter-American Tropical Tuna Commission, 1900s–1990s*.
Wapner, Paul (1995), 'Politics Beyond the State: Environmental Activism and World Civic Politics', 47(3) *World Politics* 311.
Weissbrodt, David (2005), 'Business and Human Rights', 74 *University of Cincinnati Law Review* 55.
Wellens, Karel and Gustaaf Borchardt (1989), 'Soft Law in European Community Law', 14 *European Law Review* 267.
Wendt, Alexander (1987), 'The Agent-structure Problem in International Relations Theory', 41 *International Organization* 335.
Wendt, Alexander (1992), 'Anarchy is what States Make of it: The Social Construction of Power Politics', 46 *International Organization* 391.
Wendt, Alexander (1994), 'Collective Identity Formation and the International State', 88(2) *American Political Science Review* 384.

Wendt, Alexander (1995), 'Constructing International Politics', 20 *International Security* 71.
Wendt, Alexander (1998), 'On Constitution and Causation in International Relations', 24(5) *Review of International Studies* 101.
Wendt, Alexander (1999), *Social Theory of International Politics.*
Wendt, Alexander (2001), 'Driving with the Rearview Mirror: On the Rational Science of Institutional Design', 55(4) *International Organization* 101.
Wenger, Etienne (1999), *Communities of Practice: Learning, Meaning, and Identity.*
Wenger, Etienne, Richard McDermott and William M. Snyder (2002), *A Guide to Managing Knowledge: Cultivating Communities of Practice.*
Werner, Wouter (2001), 'Securitization and Judicial Review: A Semiotic Perspective on the Relation Between the Security Council and International Judicial Bodies', 14 *International Journal for the Semiotics of Law* 345.
Wessel, Jared (2006), 'Judicial Policy-Making at the International Criminal Court: An Institutional Guide to Analyzing International Adjudication' 44 *Columbia Journal of Transnational Law* 377.
Westfield, Elisa (2002), 'Globalization, Governance, and Multinational Enterprise Responsibility: Corporate Codes of Conduct in the 21st Century', 42 *Vanderbilt Journal of International Law* 1075.
Weyland, Kurt (2002), 'The Diffusion of Innovations: A Theoretical Analysis', Paper presented at the annual meeting of the American Political Science Association, Boston, 28 August.
White, Allen (2006), 'Why we need global standards for corporate disclosure', 69(3) *Law and Contemporary Problems* 167.
Wiener, Antje (2004), 'Contested Compliance: Interventions on the Normative Structure of World Politics', 10(2) *European Journal of International Relations* 189.
Wiener, Antje (2007), 'The Dual Quality of Norms and Governance Beyond the State: Sociological and Normative Approaches to "Interaction"', 10(1) *Critical Review of International Social and Political Philosophy* 47.
Wiener, Jonathan (1999), 'Global Environmental Regulation: Instrument Choice in Legal Context', 108 *Yale Law Journal* 677.
Wildhaber, Luzius (1980), 'Some Aspects of the Transnational Corporation in International Law', 27 *Netherlands International Law Review* 79.
Williams, Colin and Andrew Millington (2004), 'The Diverse and Contested Meanings of Sustainable Development', 170(2) *The Geographical Journal* 99.
Williamson, Oliver (1975), *Markets and Hierarchies: Analysis and Antitrust Implications.*

Winter, Gerd (ed.) (2006), *Multilevel Governance of Global Environmental Change: Perspectives from Science, Sociology and the Law.*

Winter, Gerd (ed.) (2006), 'The Legal Nature of Environmental Principles in International, EU and Exemplary National Law' in Gerd Winter (ed.), *Multilevel Governance of Global Environmental Change: Perspectives from Science, Sociology and the Law*, p.587.

Wippman, David (2004), 'The International Criminal Court' in Christian Reus-Smit (ed.), *The Politics of International Law*, p. 151.

Wirth, David (1995), 'The Rio Declaration on Environment and Development: Two Steps Forward and One Back, or Vice Versa', 29 *Georgia Law Review* 599.

Witteveen, Willem (2005), 'Turning to communication in the study of legislation' in Nicolle Zeegers, Willem Witteveen and Bart van Klink (eds), *Social and Symbolic Effects of Legislation Under the Rule of Law*, p.17.

Wouters, Patricia (ed.), *International Water Law: Selected Writings of Professor Charles B Bourne.*

Wright, Christopher and Alexis Rwabizambuga (2006), 'Institutional Pressures, Corporate Reputation, and Voluntary Codes of Conduct: An Examination of the Equator Principles', 111(1) *Business and Society Review* 89.

Young, Oran (2004), 'Institutions and the Growth of Knowledge: Evidence from International Environmental Regimes', 4 *International Environmental Agreements: Politics, Law and Economics* 215.

Zeegers, Nicolle, Willem Witteveen and Bart van Klink (eds) (2005), *Social and Symbolic Effects of Legislation Under the Rule of Law.*

Zimbardo, Philip and Michael Leippe (1991), *The Psychology of Attitude Change and Social Influence.*

Documents of international institutions

Business Leaders' Initiative on Human Rights, United Nations Global Compact, and Office of the High Commissioner for Human Rights (2005), *A Guide for Integrating Human Rights Into Business Management*, United Nations Global Compact, www.unglobalcompact.org/docs/issues_doc/human_rights/guide_hr.pdf at 15 May, 2008.

Commission on Global Governance, The (1995), *Our Global Neighborhood: The Report of the Commission on Global Governance*.

Global Compact Office (2003), *Experiences in Management for Sustainability*, United Nations Global Compact, www.unglobalcompact.org/docs/news_events/8.1/exp_man.pdf, accessed on 15 May, 2008.

Global Compact Office (2003), *From Principles to Practice*, United Nations Global Compact, www.unglobalcompact.org/docs/news_events/8.1/princ_prac.pdf, accessed on 15 May, 2008.

Global Compact Office (2005), *The United Nations Global Compact: Advancing Corporate Citizenship*, United Nations Global Compact, www.unglobalcompact.org/docs/about_the_gc/2.0.2.pdf, accessed on 15 May, 2008.

Global Compact Office (2005), *The Global Compact Communication on Progress: A Status Report and Recommendations for Improvements*, United Nations Global Compact, www.unglobalcompact.org/docs/news_events/9.1_news_archives/2005_07_15/statrep_cop2.pdf, accessed on 15 May, 2008.

Global Compact Office (2006), *Leading the Way in Communication on Progress*, United Nations Global Compact, www.unglobalcompact.org/docs/communication_on_progress/4.3/leading_the_way.pdf, accessed on 15 May, 2008.

Global Compact Office (2007), *UN Global Compact Annual Review 2007 Leaders' Summit*, United Nations Global Compact, www.unglobalcompact.org/docs/news_events/8.1/GCAnnualReview2007.pdf, accessed on 15 May, 2008.

Global Compact Office (2008), *Policy for 'Communication on Progress'*, United Nations Global Compact, www.unglobalcompact.org/docs/communication_on_progress/COP_Policy.pdf, accessed on 15 May, 2008.

Global Compact Office and the Barcelona Centre for the Support of the

Global Compact on the Local Networks (2007), *Local Network Report*, United Nations Global Compact, www.unglobalcompact.org/docs/news_events/8.1/LNReport_FINAL.pdf, accessed on 15 May, 2008.

Intergovernmental Panel on Climate Change (2005), *Carbon Dioxide Capture and Storage*.

KPMG, Royal Danish Ministry of Foreign Affairs and the United Nations Development Programme (2005), *Implementing the UN Global Compact: A Booklet for Inspiration*, www.unglobalcompact.org/docs/news_events/8.1/dk_book_e.pdf, accessed on15 May, 2008.

McKinsey & Company (2004), *Assessing the Global Compact's Impact*, United Nations Global Compact, www.unglobalcompact.org/docs/news_events/9.1_news_archives/2004_06_09/imp_ass.pdf, accessed on 15 May, 2008.

McKinsey & Company (2007), *Shaping the New Rules of Competition: UN Global Compact Participant Mirror*, www.unglobalcompact.org/docs/summit2007/mckinsey_embargoed_until020707.pdf, accessed on 15 May, 2008.

Office for the London Dumping Convention at the International Maritime Organization (1991), *The London Convention: The First Decade and Beyond*.

Organisation for Economic Cooperation & Development (1999), *Handbook of Incentive Measures for Biodiversity: Design and Implementation*.

OSPAR Commission for the Protection of the Marine Environment of the North-East Atlantic, 'Placement of Carbon Dioxide in Sub-Sea Geological Structures', www.ospar.org/eng/html/welcome.html, accessed on15 May, 2008.

United Nations Department for Economic and Social Affairs, Division for Sustainable Development (2005), *International Rivers and Lakes Newsletter*, No. 43, www.un.org/esa/sustdev/sdissues/water/rivers_lakes_newsletter.html, accessed on 15 May, 2008

Index